Sustainability of Temperate Forests

by
Roger A. Sedjo
Alberto Goetzl
and
Steverson O. Moffat

Resources for the Future
Washington, DC

Printed in the United States of America

Published by Resources for the Future
1616 P Street, NW, Washington, DC 20036–1400

Library of Congress Cataloging-in-Publication Data

Sedjo, Roger A.
Sustainability of temperate forests / by Roger A. Sedjo, Alberto Goetzl, and Steverson O. Moffat
    p.   cm.
    Includes bibliographical references and index.
    ISBN 0–915707–98–5

    1. Sustainable forestry. I. Goetzl, Alberto. II. Moffat, Steverson O., 1965– . III. Title.
SD387.S87S44   1998
333.75′15′0912—dc21   98–28796
                                                           CIP

The paper in this book meets the guidelines for permanence and durability of the Committee on Production Guidelines for Book Longevity of the Council on Library Resources.

This book is a product of the Division of Energy and Natural Resources at Resources for the Future, Michael A. Toman, director. It was copyedited by Flora Szatkowski and typeset in Palatino by Betsy Kulamer; its cover was designed by AURAS Design.

About
Resources for the Future

R esources for the Future is an independent nonprofit organization engaged in research and public education with issues con-
cerning natural resources and the environment. Established in
1952, RFF provides knowledge that will help people to make better
decisions about the conservation and use of such resources and the
preservation of environmental quality.

RFF has pioneered the extension and sharpening of methods of
economic analysis to meet the special needs of the fields of natural
resources and the environment. Its scholars analyze issues involving forests, water, energy, minerals, transportation, sustainable
development, and air pollution. They also examine, from the perspectives of economics and other disciplines, such topics as government regulation, risk, ecosystems and biodiversity, climate, Super-
fund, technology, and outer space.

Through the work of its scholars, RFF provides independent
analysis to decisionmakers and the public. It publishes the findings
of their research as books and in other formats, and communicates
their work through conferences, seminars, workshops, and brief-
ings. In serving as a source of new ideas and as an honest broker on
matters of policy and governance, RFF is committed to elevating
the public debate about natural resources and the environment.

Contents

Preface

Until a few years ago, most discussions of sustainability of the world's forests centered on tropical rain forests, because these forests were experiencing rapid deforestation. More recently, discussion is turning towards the condition of temperate forests. This shift was driven partly by the concerns of the developing countries, which felt that they were being singled out for their forestry behavior, and by concerns that many temperate countries were practicing questionable forestry management.

This book summarizes some of the key recent developments relating to policies and institutions that are of significance to the availability and competitiveness of products from temperate forested countries. Our investigation focused on changes in recent years, largely since the UNCED meeting in Rio in 1992. It attempts to provide a broad overview of the forces acting on temperate forests at a number of levels: governmental, international, national, private industrial, and nongovernmental. In Appendix A, case studies are presented for eight countries: Canada, Chile, Germany, Finland, France, New Zealand, Sweden, and the United States. These countries were chosen because they are each important wood producers, as well as major wood exporters or importers.

The authors would like to acknowledge the assistance of a large number of individuals in a number of countries. These include Ake Barklund, Nordic Forest Certification Project; Lars Lonnstedt, University of Uppsala, Sweden; Birger Solberg and Eeva Hellstrom, European Forest Institute; Bjorne Hagglund of Stora Skog, Sweden;

Ola Sallnas, Alnarp, Sweden; Matti Palo, Finnish Forest Research Institute; I. Tikkanen, University of Helsinki; John Spears, World Commission on Forestry; Claire Hubert and colleagues, Association Foret-Cellulose, France; Bernard Chevalier, Des Eaux et Forets, France; Heiner Ollman, Federal Research Center for Forestry and Forest Products, Germany; Arno Früwald, University of Hamburg, Germany; Ken Shirley, Member of Parliament, New Zealand; Colin McKenzie and James Griffiths, New Zealand Forest Industries Council; Gill Chappell and Murray Parrish, Carter Holt Harvey, New Zealand; Don Wijewardana, Ministry of Forestry, New Zealand; Wink Sutton, Fletcher Challenge, New Zealand, Gordon Hosking, G. Tattersall Smith, and Gerald Horgan, New Zealand Forest Research Institute; Peter J. Berg, Rayonier New Zealand; John Schrider, FORME Consulting, New Zealand; Gonzalo L. Paredes, Universidad Austral de Chile; Bertram Husch and Alfredo Unda, INFORA, Chile; Eduardo Morales, Fundacion Chile; Antonio Grass, CONAF, Chile; Don Taylor and Carlton Owen, Champion International; John McMahon, Weyerhauser Company; William Caferata, Neil Brett-Davies, and others, MacMillan Bloedel, Vancouver, B.C.; John Reid, Pacific Forest Trust; Robert Hendricks and Mary Columbe, U.S. Forest Service; Scott Berg and John Heissenbuttel, American Forest and Paper Association; Richard Donavan, Smart Wood; Michael Jenkins, MacArthur Foundation; Bill Mankin, Global Forest Policy Project; Chris Elliot, World Wildlife Foundation; Francis Sullivan, Forest Stewardship Council; I.J. Bourke and K.H. Schmincke, Food and Agriculture Organization; an anonymous reviewer; and many others who influenced this piece while in progress. Nevertheless, the authors bear full responsibility for the final product.

Bringing this product into published form requires us to note also the contributions of several people at Resources for the Future. Kay Murphy provided her usual exemplary assistance in preparing all the versions of the manuscript. Useful suggestions came from Mike Toman, Paul Portney, Rich Getrich, Chris Kelaher, Eric Wurzbacher, and Kelly See that helped the authors fine-tune parts of the manuscript into the book you hold before you.

Finally, we are thankful for the financial and moral support for this study, which came from the U.S. Forest Service and the Foreign Agricultural Service, as well as through the general support funding of the Forest Economics and Policy Program from the American Forest and Paper Association and others in the forest industry, as well as RFF.

Roger A. Sedjo, Resources for the Future
Alberto Goetzl, Consultant
Steverson O. Moffat, North Carolina State University

1

Sustainable Forestry

An Overview

In the years since the June 1992 "Earth Summit"—more formally, the United Nations Conference on Environment and Development (UNCED)—in Rio de Janeiro, Brazil, forest sustainability has been the focus of a number of international and domestic government actions, as well as private efforts. These actions include numerous dialogues, some nonbinding international agreements, a host of policy changes within countries, and the creation of several strategies to promote sustainable forestry through management standards, professional associations, and third-party certification. Although UNCED was convened largely because of global concerns about tropical deforestation, much of the attention since Rio has been focused on temperate forested countries. This was done in part because it was recognized that the temperate countries, as well as the tropical, also had problems with their forests. These countries account for over 80% of the world's industrial roundwood production (that is, raw wood) and global trade in wood and paper products.

This study has three basic objectives. First, it examines international and national efforts to address forest sustainability in temperate and boreal countries, and discusses the key developments affecting sustainable forestry and international trade. Second, it assesses some of the considerations and consequences likely to flow from this process, especially on how the various countries and forest industries are adapting to the new laws, policies, and regulations, in the context of new production and market realities, with a special emphasis on the general influences on sustainable forestry and international trade patterns. Third, it compiles information for a large number of countries on forest practices, laws, and regulations and examines how various changes in international and domestic institutions, including newly instituted laws and regulations, are affecting land management decisions and outputs. There is a particular focus on eight temperate forested countries (Canada, Chile, Finland, France, Germany, New Zealand, Sweden, and the United States) and brief case studies are presented.

The international, national, and private approaches to sustainable forestry offer a diverse variety of perspectives on sustainable forestry. At one extreme is the view that current practices are sustainable, a view based largely upon the data that show temperate forest areas have been increasing in many countries in which forestry has been practiced (in one way or another) for nearly a thousand years. At another extreme is the view that there are many indicators that practices in these forests may not be sustainable and that careful third-party monitoring of on-the-ground forestry practices is required to ensure that sustainable forest management is practiced. Between these extremes are a number of alternative approaches. These include those represented by the multinational Montreal and Helsinki processes, which focus on national level approaches to promote sustainable forestry; the International Standards Organization's Environmental Management Standard, which is focusing on developing standards for the environmental management systems within firms; industry-based sustainable management guidelines and standards such as the Sustainable Forestry Initiative in the United States and those of the Canadian Standards Association (CSA) in Canada; and finally an externally based set of sustainable forestry practices standards involving third-party on-the-ground certification.

These alternative approaches are clearly not all apolitical. The interests of the various groups are diverse. Countries have concerns about the effects of the various approaches on the competitive position of their forest industry in the world market as well as the impact on the quality of their forests. An economic analyst may wonder whether the costs of these various approaches are justified by their social benefits, especially if the positive view of the condition of temperate forests is accepted. A civil libertarian may ponder the implications of the various schemes on the trade-off between private and public rights. Industrial firms may judge the various approaches by their anticipated impact on the firm and on its competitors, both domestic and international. Environmental groups may view this whole issue as a vehicle to improve the forests, or as a way to increase their membership, revenues, and influence, or both. The process we will be describing in this volume is replete with various special interest agendas.

Nevertheless, virtually all of the temperate forested countries, including the United States, have participated in international discussions and have entered into nonbinding agreements related to sustainable forestry. Many countries have also instituted changes in their regulatory and institutional oversight of forest management activities in recent years. Nongovernmental programs, including forest certification, are also being promoted. All of these developments potentially have significant implications for competitiveness among suppliers of temperate forest products in world markets: they also potentially have significant implications for the long-term condition of the forests.

2

History and Evolution of Sustainable Forestry Concepts

Humans have grappled with the challenge of maintaining forest outputs of one kind or another throughout much of history. The science and practice of forestry were born of sustainability concerns and continue to evolve as human attitudes toward forests change. The international, national, and nongovernmental private approaches to sustainability that have been proposed in the years since the 1992 Earth Summit are little understood by the vast majority of forest product consumers. All the approaches stand to raise costs to a greater or lesser degree; this book speculates as to whether particular countries or types of producers are likely to gain market share by adopting a particular approach. Both political pressure and market pressure will require producers to increasingly pay attention to forest sustainability issues, and thus to the origins of these issues. This chapter briefly describes history, development, and current status of international efforts and initiatives in sustainable forestry.

EVOLVING CONCEPTS OF FOREST SUSTAINABILITY

Ideas of sustainability as related to forests have a long history, dating back to the Old Testament and ancient Chinese writings (Menzies 1992). Forests have always been relied upon for food, game,

5

recreation, and hunting, as well as for fuel and construction materials. In the Middle Ages, Europeans became concerned with managing forests to maintain a continuous supply of forest outputs even while agriculture was expanding. However, while forests were valued for their outputs, they were also an impediment to the expansion of agriculture and grazing. Thus, in many parts of the world, forests were gradually cleared by intentional conversion for pasture and cropping, encroachment by browsing animals, and the utilization of wood for fuel and construction. Concerns about forest use and depletion gave rise to the science of forestry in the eighteenth and nineteenth centuries. Forest management incorporated concepts of sustainability in terms of continuous production of wood fiber, and managed forest stands were viewed as being capable of producing a sustained yield of timber commodities in perpetuity.

By the late nineteenth century and early twentieth century, crop or livestock production was no longer economically advantageous for many areas in temperate counties. Consequently, such areas began to revert back into forests. However, new issues began to emerge which focused attention on environmental aspects of forests related to water quality, wildlife habitat, aesthetics, recreation, and other noncommodity outputs. With the rise of these environmental concerns, the existence and maintenance of forests began to take on additional significance. More recently, sustainability has become the centerpiece of global discussions on forests, and efforts to ensure forest sustainability have begun to influence land management decisions and potentially pose new challenges in international trade. As both domestic and international environmental concerns escalated, a strong interest in implementing some type of sustainable forestry has emerged.

The concept of what constitutes sustainable forestry is dynamic and evolving. There is no clear definition of sustainable forestry although there is a general understanding that sustainability refers to balancing environmental, social, and economic needs in such a way as to accommodate those same needs in the future. The Brundtland Commission report best captured the modern concept of sustainability when it described sustainable development as "development that meets the needs of the present without compromising the ability of future generations to meet their own needs" (WCED 1987). Likewise, Flasche (1997) was probably correct when he characterized today's sustainable forestry as more a philosophy of how forests should be cared for than a definable condition of the forests or a set of acceptable management practices. Clearly, much of the complexity involved in the concept of sustainability is due to the necessary consideration of an array of economic, social, and philosophical management goals.

Some specific forestry policies aimed at sustainability essentially attempt to set a standard for maintaining and monitoring sustainable forest practices. Forest certification, for example, is a process of assuring that a forest is being managed sustainably. Yet even advocates of certification acknowledge that "scientific data do not yet support a single consensus on definition of biological sustainability, especially given regional variations in ecology; the same is true for socioeconomic sustainability" (Heaton and Donovan 1997, 55). Furthermore, some have argued that although certification may provide useful incentives, by itself it is unlikely to ensure sustainable forest management (Kiekens 1995). For example, most tropical deforestation is generated by factors other than commercial logging, primarily conversion of forestlands (FAO 1987).

Even if a forestry management policy such as certification cannot by itself ensure sustainability, it can be argued that certification could be a catalyst for management changes (Viana 1997). In most temperate forested countries and in many tropical countries, forest policies have been changing to address forest sustainability issues. Forest sustainability is now viewed as a function of several factors, including landowner and societal objectives for the forest, the size of the forest area or landscape (a continuum from forest management unit or stand to watershed, region, nation, or global area), past and current forest ecosystem conditions, and time. Since these are all dynamic variables, defining sustainability—or, more appropriately, reaching consensus on sustainability—is a difficult task. Yet despite this difficulty, achieving forest sustainability given some accepted definition has become a de facto goal of most governments and international agencies involved in natural resource management.

MAJOR INTERNATIONAL DEVELOPMENTS ON
SUSTAINABLE FORESTRY

Attention to global forestry and forest sustainability at the international level has clearly been elevated in recent years. President George Bush raised the issue during a June 1990 meeting of the G-7 leaders in Houston, Texas. At that meeting, the leaders of the industrialized nations issued a statement citing the need to reverse the loss of the world's forests. They also called for the negotiation of an international convention on forests to help curb deforestation, protect biological diversity, stimulate positive forestry actions, and address other threats to the world's forests (Drake 1995). Global forestry issues were again raised at subsequent meetings of world leaders in anticipation of the United Nations Conference on Environment and Development (UNCED) held in Rio de Janeiro in June 1992. The attention given to global forestry issues by U.S. and other world leaders during the 1990s has been historically unprecedented.

Figure 1 schematically depicts a number of developments related to sustainable forestry programs and initiatives. The major processes are described below.

UNCED: The Earth Summit

The United Nations Conference on Environment and Development (UNCED), or Earth Summit, was several years in the making. With respect to forestry issues, one of the main drivers for the conference was concern over tropical deforestation. A number of publications throughout the 1970s and 1980s pointed to alarming rates of tropical deforestation. In the early 1980s, the Food and Agriculture Organization of the United Nations (FAO) completed a global assessment of the rates of tropical deforestation. Although the FAO estimates turned out to be considerably lower than some of the wider claims, the rate of tropical deforestation was still substantial, about eleven million hectares per year for the late 1970s, high enough to generate appropriate concerns.

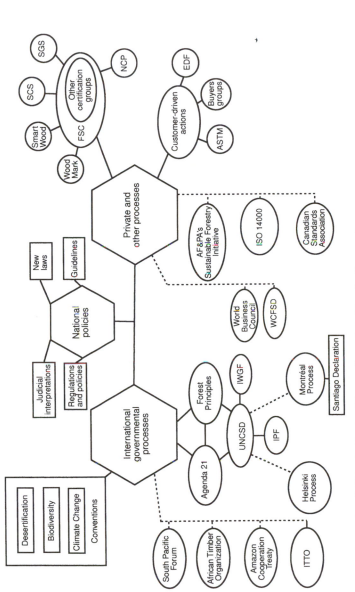

Figure 1. Forest Sustainability Efforts Since the 1992 Earth Summit.

Source: This chart is adapted from a memo prepared by C.N. Owen.

Note: AF&PA, American Forest and Paper Association; ASTM, American Society for the Testing of Materials; EDF, Environmental Defense Fund ; FSC, Forest Stewardship Council; IPF, Intergovernmental Panel on Forests ; ITTO, International Tropical Timber Organization; IWGF, Intergovernmental Working Group on Forests; SCS, Scientific Certification Systems; SGS, Societe General de Surveillance; UNCSD, UN Commission on Sustainable Development; WCFSD, World Commission on Forests and Sustainable Development.

In the dialogue leading up to UNCED, it is worth noting that few serious global concerns were expressed about the overall condition of the temperate and boreal forests, though concerns were raised for various localized problems. Parts of Europe, for example, were believed to have been experiencing forest dieback, which some attributed to acid rain. This turned out to be much less of a problem than initially thought. In fact, subsequent evidence suggests that, contrary to widespread concerns of dieback of European forests in the 1980s, European forests in all regions experienced high rates of biological growth throughout the decade of the 1980s (Kuusela 1994). In the United States concern was also expressed with regard to the management of the old-growth Douglas fir forest in the Pacific Northwest and to large clearcutting activities that were occurring in parts of the United States, Canada, and the Soviet Union. However, even where harvests were high, there was little warranted concern as to the continuation of the forest. Temperate forests, after being harvested, almost always regenerated back into viable forests and, on balance, were not being displaced by nonforest land uses.

A total of 178 governments were represented at UNCED, and a contentious debate revolved around forest issues. While some temperate countries took a cautious approach, tropical countries objected strongly to any binding proposals on forests that might infringe on their sovereignty. Instead, nations agreed to a broad agenda to address environmental and development issues. UNCED ultimately produced four documents that related in whole or in part to forests:

- Agenda 21, a global environmental action plan
- A framework convention on global climate change
- A convention on biological diversity
- Statement of Principles on Forests

The Statement of Principles on Forests reflects a global consensus on a set of nonbinding principles of management, conservation, and sustainable development of all types of forests. The discussions and documents coming out of the Earth Summit also broached some specific management issues, such as the issue of certification and ecolabeling. For example, Agenda 21 encourages "expansion of environmental labeling and other environmentally related product information programs designed to assist consumers to make informed choices." Thus, the agenda and resulting documents are flexible enough to represent a loose consensus, while still proposing some specific policy options.

At UNCED, world leaders also agreed to form the U.N. Commission for Sustainable Development (UNCSD), which was established to follow up on the initiatives coming out of UNCED. UNCSD is comprised of fifty-three representatives elected from U.N. member countries. UNCSD, in turn, established the Intergovernmental Panel on Forests (IPF) to provide a forum for addressing the processes of moving to some acceptable agreements on sustainable forestry. Independent from UNCSD, Malaysia and Canada jointly sponsored an Intergovernmental Working Group on Forests (IWGF) to provide a forum for resolving North/South, or tropical versus temperate country, differences. The final report of the IWGF was incorporated into later discussions at the IPF.

One of the major outcomes of the Earth Summit was the industrial worlds' agreement that sustainable forestry should be practiced by all countries, both tropical and temperate. In the process, the focus ironically shifted from the tropical countries, which had been the source of deforestation concerns of the 1980s, to the industrial countries, where little serious concern had existed about deforestation. As the tempo of attention about temperate forest sustainability has increased in the post-UNCED period, the earlier problem of tropical deforestation seems to have receded into the background.

In response to international and domestic concerns, various countries have made changes to their domestic laws and policies to improve water quality, protect biological diversity, and implement less intensive silvicultural treatments. Some of these policies are regulatory in nature, while others rely on taxation or other incentives. (See Table 1 for a summary of these policies and regulations in the eight nations of this study.) In all of the temperate forested countries examined, significant revisions have been made over the past several years in the legal and/or institutional framework deal-

Table 1. Existing Forest Policy/Regulation in Selected Temperate Forested Countries.

Country	Relevant laws	Major provisions	Comments
Canada	Forest Practices Code (1993)	Regulates silvicultural activities to protect riparian areas and enhance biodiversity	According to some, new regulations reduce harvests per unit area by as much as 35% and reduce total area available to harvest by 24%. *Certification/standards*: Canada has developed performance standards for forest practices through the Canadian Standards Association.
Chile	Forest Decree 701 (dates back to 1974, but was extended in 1994 and 1997); New regulations were recently promulgated under Forest Law of 1931; Environmental Law was enacted in 1993; Native Forest Law was enacted in 1997.	Laws and regulations require forest management plans in order to qualify for generous subsidy and preferential tax treatment; new regulations require that greater emphasis on environmental amenities be included in plans; native forests will be subject to stricter regulation on harvesting and conversion.	New environmental law requires environmental impact statement for major projects; Supreme court ruled against large forest development project in southern Chile that was designed to meet sustainability objectives. *Certification/standards*: Issues are being explored, but no official or consistent approach is being considered; Chile is working on ways to monitor criteria and indicators.
Finland	Forest and Park Service Act (1994); government issued new policy, Environmental Guidelines for Forestry in Finland (1994), and enacted revised Forest Law and new Environmental Law.	New forest policy places biodiversity on an equal footing with timber production; law requires that public lands be managed more extensively; all landowners are required to conduct an inventory of plant and animal species.	New regulations will require some lands to be taken out of timber production. *Certification/standards*: Finnish certification system recently announced by government, industry, and landowner associations; it is not approved by the Forest Stewardship Council; Finland is cooperating on the Nordic certification project.
France	New forestry law under consideration; system of regional and local regulations governed under a Forest Code dating back to 1827.	Detailed forest plans required for holdings of 25 hectares or larger; law probates conversion of inherited forest land for 30 years.	Most forest regulation is local. *Certification/standards*: France is developing criteria and indicators measures.

Germany	Forest regulation primarily state and local; Federal Forestry Act dates back to 1975, but only provides framework; Nature Protection Law recently enacted;	Forest regulations include prohibition against clear cutting; forest management plans required of holdings greater than 30 hectares; subsidies favor hardwood management.	Common practice is for long rotations; government policy encourages conversion of conifer stands to native hardwood species. *Certification/standards*: Most German forest interests cite history of German forestry as proof of sustainability.
New Zealand	Resource Management Act (1991)	The RMA is a national land-use planning law; regional councils can require permits for activities that have environmental effects.	*Forest Accord* between industry and conservation groups has eliminated commercial harvests from native forests; industry has adopted *Forest Principles*. *Certification/standards*: New Zealand industry favors ISO process model for sustainability.
Sweden	Forestry Act (1993); Nature Conservation Act amendments.	The greater emphasis is on biodiversity through protection of riparian areas, restrictions on planting of exotic species, and regulations on age distribution (no more than 50% of forest younger than 20 years).	New laws have eliminated subsidies for forest planting. *Certification/standards*: Major industry firms have reached agreement on a certification system approved by the Forest Stewardship Council (FSC); the major forest cooperatives, however, are not in agreement with FSC on certification; Sweden is cooperating on the Nordic certification project.
United States	Policy environment characterized by court interpretations of existing laws including Endangered Species Act, National Environmental Policy Act, and Clean Water Act; many states have forest practices acts or other regulatory programs.	Protection of threatened and endangered species takes precedence over other management objectives; protection for riparian areas regulated or subject to voluntary best management practices (BMPs).	The United States is only country with separate and specific legal authority to list and protect species; public land harvests have been dramatically reduced *Certification/standards*: A voluntary industry program (Sustainable Forestry Initiative) is being implemented; America Tree Farm System has certified 95 million acres; still evolving a combination of systems and performance approach.

ing with forest matters. In addition, private efforts have been initiated. These include the American Forest and Paper Association's Sustainable Forestry Initiative and a forest industry-specific standard by the Canadian Standards Association. Two nongovernmental, nonindustry private organizations are the World Commission on Forests and Sustainable Development, which is obtaining comments and inputs from interested parties around the world, and the Forest Stewardship Council, which is the major nongovernmental organization working to develop standards for sustainable forestry practices.

Developing Criteria and Indicators

Following the Earth Summit, many varied international activities were organized around forest sustainability issues. A number of separate but related discussions were held by different groups of countries to follow up on the UNCED initiatives. Generally, these meetings were geared to defining what sustainability meant to forest management and to developing criteria and indicators for sustainable forestry. In 1993, the majority of European nations convened the 1993 Ministerial Conference of the Protection of Forests in Europe and endorsed the Helsinki Process, which defined six criteria for characterizing sustainable forests.

For non-European temperate forest countries, a Working Group on Criteria and Indicators for the Conservation and Sustainable Management of Boreal and Temperate Forests was formed in 1994. A series of discussions, which became known as the Montreal Process, led to the Santiago Declaration of February 1995. The Santiago Declaration, like the Helsinki Process, contains a set of criteria and indicators endorsed by the participating countries (Argentina, Australia, Canada, Chile, China, Japan, Mexico, New Zealand, South Korea, Russia, the United States, and Uruguay). The seven criteria are said to characterize sustainable forest management. They cover biodiversity conservation, ecosystem productivity, ecosystem health and vitality, soil and water conservation, global carbon cycles, multiple socioeconomic benefits, and legal/policy/institutional frameworks. These were intended to provide a common understanding of what is meant by sustainable forest management and provide a common framework for describing, assessing, and evaluating a country's progress toward sustainability *at the national level*. Thus, they help provide an international reference for policymakers in the formulation of national policies and a basis for international cooperation aimed at supporting sustainable forest management.

Related to the overall policy framework that is designed to facilitate the conservation and management of forests of a country is Criterion 7 and its associated indicators. Criterion 7 includes twenty indicators that address the economic, institutional, and legal frameworks for forest conservation and sustainable management. Indicators of these criteria include: well-defined property rights; the extent of periodic forest-related planning; the use of best-practice codes for forest management; the provision of management for the conservation of special environmental, cultural, social and scientific values; adequate infrastructure to facilitate the support of forest products and management; the enforcement of laws, regulations, and guidelines; investment and tax policies; nondiscriminatory trade policies for forest products; the ability to measure and monitor changes in conservation and sustained management; and the capacity to conduct and apply research. (Appendix B of this book describes Criterion 7 in greater detail.)

Although *criteria* are broad descriptions of economic, social, and ecological conditions that can collectively contribute to sustainability, *indicators* are measures or descriptions of the criteria that can be monitored over time. Criteria and indicators developed in the Montreal Process are intended for use at the national level and are purposely not site-specific measures. Signatories to the Santiago Declaration met in Korea in July 1997 to present and discuss "First Approximation" reports on criteria and indicators as applied in each country. The outline of criteria and indicators developed through both the Montreal and Helsinki processes included social and economic indicators related to forests and forest production activities. As such, they recognize current concepts of sustainable forestry beyond purely biological and physical dimensions by linking human welfare to forest management.

Although not detailed in this book, tropical countries have also focused attention on sustainable forestry issues. Criteria and indicators were first developed through the International Tropical Timber Organization (ITTO) (1990) and the Tarapoto Proposal for the Amazonian Forests (1995). ITTO has been working to implement its Objective Year 2000 Program whereby all ITTO-producing countries would trade in timber from sustainable sources. Major European importing countries have pressed hard for the ITTO to implement its sustainable forestry program.

Intergovernmental Panel on Forests

To adequately address the increasingly complex issues dealing with forest management, UNCSD established the Intergovernmental Panel on Forests (IPF) in April 1995. While endorsing the criteria and indicators approach, the IPF has attempted to refine and sort out conflicting views on a number of global forest issues. One of these is the need for and desirability of international forest convention. Many environmental groups and many countries, including the United States, now oppose a binding international convention on forests. Some organizations believe that, because of the breadth of disagreement on issues, negotiations toward a convention would be exhaustive and would not result in a substantive improvement over existing international agreements. They believe that existing conventions on biological diversity and climate change may have a greater impact. Many in the United States also believe that it would be difficult to garner political support for an international convention on forests. A number of environmental nongovernmental organizations (NGOs) have been advocating a forestry protocol to be included within the Biological Diversity Convention instead, but to date there has not been any consensus.

The IPF held four meetings. While numerous issues were addressed, the IPF was unable to reach a consensus on negotiating an international agreement. The panel deferred the issue to the full World Commission on Sustainable Development that met in April 1998. Similarly, the UNCSD could not reach a consensus and decided to defer the issue to the full U.N. General Assembly that met in June 1998 to discuss five years of progress since the Earth Summit. The options ranged from continuing forestry discussions within existing international forums to establishing a committee to pursue a legally binding forestry convention. The IPF reaffirmed that any forest-related certification or other initiative should be nondiscriminatory and transparent, and ensure open access by not imposing barriers to trade.

Some environmental NGOs are pushing to include more timber species in the Convention on the International Trade of Endangered Species (CITES), thus extending prohibitions against trade of additional timber products. Extending CITES coverage to other species might also regulate trade in "like species or products" where it is difficult to differentiate between listed and unlisted species.

World Commission on Forests and Sustainable Development

The World Commission on Forests and Sustainable Development (WCFSD) is a private, nonprofit organization created for a limited time period, whose purpose is to "search out opportunities for mutual gain among stakeholder groups with emphasis on consensus building at all levels of society on peaceful and constructive dialogue." In this role the WCFSD has undertaken a number of "public hearings" in various regions around the world to identify policy and institutional reforms that will ease the transition as forest management objectives are adjusted to give increased emphasis to protection goals. To some degree the WCFSD can be viewed as a private NGO, paralleling the IPF. However, it has not issued its final report as of this writing.

3

Alternative Approaches to Defining Sustainable Forestry

While there is a general consensus that sustainable forestry in some form or another should be practiced, there is clearly less agreement regarding what are considered to be sustainable forest conditions and how to ensure that these are achieved.

At one extreme are those who maintain that sustainable forestry is already being practiced in much of the developed world. It has been recognized among land historians and land use experts that the temperate and boreal forests, which dominate this region, have been stable or expanding for decades, if not centuries. It is not disputed that the net deforestation that is taking place in the world is almost entirely in the tropics. It is also widely agreed that most of the conversion of temperate forests to other uses took place in past centuries, and that temperate forests have been expanding in many regions as marginal agriculture areas are abandoned and return, largely through natural regeneration, to forest (for example, see Williams 1988; Kuusela 1994). In recent decades temperate forest areas have continued to expand modestly (Korotov and Peck 1993), and net reforestation is common in much of the temperate forested world. Indeed, most of the evidence indicates that the forest volume or stocking of forests in the temperate world has been stable or increasing slightly in recent decades (Sedjo 1992).

At the other end of the spectrum are those who hold the view that sustainable forestry involves more than forest stocking. To promote sustainable forests in an ecosystem sense, they argue that third-party certification—perhaps involving product labeling to encourage participation—offers the only assurance that sustainability is being achieved. Stakeholders in some countries view third-party certification systems as an appropriate and effective means for assuring consumers that products are being manufactured in accordance with environmentally sensitive management practices. Others view third-party certification as unnecessary, overly costly, and a potentially onerous barrier to the free trade of forest products.

The various perspectives on forest sustainability can be grouped into five categories, each of which are discussed briefly in this chapter.

- Sustainability of current practices
- Criteria and indicators
- Management system-based sustainability standard (ISO 14001)
- Industry-based sustainability standards
- Sustainability requirements for third-party certification and monitoring

SUSTAINABILITY OF CURRENT PRACTICES

Proponents of the first perspective (noted above) point to the fact that the area in forests and the forest inventory in the temperate countries have stabilized, suggesting that current practices are sustainable by traditional definitions. This view maintains that the continuation of forest cover on the land over very long periods of time, or its natural regeneration after alternative land uses, attests to the inherent sustainability of the forest. Changes in species mixes are not viewed as necessarily damaging because forests are inherently dynamic and ecosystem services can be generated by a host of different tree and plant combinations (Grime 1997). The argument that contemporary forest practices are much more environmentally benign than some of those practiced in the past also lends support to this view.

A European government official has maintained, for example, that the existence of healthy forests in Europe, after a millennium of forest practices, demonstrates the sustainability of the methods being practiced, especially since the intensity of harvesting and land clearing today is less than in times past. German foresters point to two hundred years of regulated forest management as evidence of forest sustainability.

Forest conditions in much of North America could also be cited to show the adequacy of current practices, particularly in light of the inherent resiliency of the American forest. The existence of what the U. S. Forest Service has called the South's fourth forest, or fourth rotation, attests to the resiliency of the forest and the appropriateness of the forest management systems that have been used. This view is even more relevant when it is recognized that much of the South's original forest was cleared; large periods of intervening agriculture were practiced and subsequently abandoned; and the forest was reestablished, largely through natural regeneration.

APPROPRIATE PROCEDURES ARE REQUIRED: CRITERIA AND INDICATORS

Another approach to promoting forest sustainability is characterized by development at the national level of principles and guidelines for sustainably managed forests. These guidelines are nonregulatory in nature and attempt to lay out in general terms what kinds of practices foster sustainability. Industries in New Zealand, the United States, Canada, and the Scandinavian countries subscribe to principles of sustainable forest management, which were developed internally, largely by the governments of the various countries. The international development of criteria and indicators (as described in Chapter 2) was intended to provide a common understanding of what is meant by sustainable forest management and provide a common framework for describing, assessing, and evaluating a country's progress toward sustainability *at the national level.* Thus, criteria and indicators help provide an international reference for policymakers in the formulation of national policies and a basis for international cooperation aimed at supporting sustainable forest management.

Specifically, Criterion 7 and its associated indicators are designed to facilitate the conservation and management of a country's forests. Related to the overall policy framework that is designed to facilitate the conservation and management of forests of a country, Criterion 7 addresses the legal, institutional, and economic framework for forest conservation and sustainable management (see Appendix B for further details). Its associated indicators include:

- Well-defined property rights
- Extent of periodic forest-related planning
- Use of best-practice codes for forest management
- Provision of management for the conservation of special environmental, cultural, social, and scientific values
- Adequate infrastructure to facilitate the support of forest products and management
- Enforcement of laws, regulations, and guidelines
- Investment and tax policies
- Nondiscriminatory trade policies for forest products
- Ability to measure and monitor changes in conservation and sustained management
- Capacity to conduct and apply research

The criteria and indicators approach to sustainability emphasizes the use of measures and tracking changes in those measures over time. All of the temperate forested countries have agreed, through either the Montreal Process or the Helsinki Process, to a set of criteria and indicators for sustainable forestry. This is an evolving process, but it is one that has governments developing the methods for tracking criteria and indicators. The presumption is that if a country pays attention to its land use and forest policies; develops systems at the national, regional, or state level to promote the best management practices (BMPs) appropriate to its forests; and provides suitable educational and transport systems, then the country's forests are likely to be well managed. Although not articulated in these terms, this approach seems to recognize that institutions in developed countries tend to value the environment and place a premium on well-managed forests, parks, protected areas, and other public goods. Simply put, the presence of institutions and infrastructure is viewed as providing the necessary and sufficient conditions to ensure the practice of sustainable forestry, at least in most industrial countries.

MANAGEMENT SYSTEMS APPROACH
TO SUSTAINABILITY: ISO 14001

This viewpoint maintains that sustainability can be achieved if sustainable practices are embodied in a firm's management system. This approach is best represented by the environmental management standard from the nongovernmental International Standards Organization (ISO). An ISO standard has had as its objective to provide a harmonized and globally recognized standard that does not present technical barriers to trade. For this reason, the ISO approach is an attractive one for many companies operating on a global scale.

The ISO has developed a specification for environmental management systems (EMSs), issued as ISO 14001 in September 1996. The ISO is also working to develop an international consensus on other related environmental standards for the 14000 series, including:

- Environmental auditing
- Environmental labeling
- Environmental performance evaluation
- Life-cycle assessments
- Terms and conditions

ISO 14001 provides EMS guidelines and standards that can be applied across industries. A firm can voluntarily agree to accept the standards as its goal and work towards their implementation. The standards can serve as the basis for an internal or external verification that a company can use to evaluate and improve its own internal procedures and from which it makes recommendations for improvements. Finally, the company can chose to use ISO 14001 as a basis for certification of management systems by having a third-party audit that certifies the company as meeting the ISO's guidelines and standards in its management systems and operations procedures. The generic nature of an ISO 14000-series environmental standard allows it to be applied to the production processes of products other than wood. Thus, materials competing with wood can be held to similar compliance.

An extension of the ISO type of approach is that advocated by Canada whereby the standards would be made specific to the forest industry. The advantage of this approach is that the standards applied to the forest industry would be designed specifically to the needs and characteristics of that industry. A criticism of this approach is that the ISO does not have a history of developing industry-specific procedures and systems.

In 1995, Canada and Australia proposed to the ISO technical committee that a more industry-specific approach should be applied to the forest industry. Although this proposal was withdrawn from the broader discussions, New Zealand provided a secretariat for additional study of using a more forestry-specific approach, and meetings were scheduled to further develop this approach. A consensus appears to be emerging that a separate reference document is needed to provide additional information that will assist the industry in implementing the ISO 14001 standard in the forest sector (Rotherham 1996).

INDUSTRY-BASED SUSTAINABILITY STANDARDS

The Sustainable Forestry Initiative (SFI) of the American Forest and Paper Association and other industry-initiated standards often have aspects somewhat different from those of the ISO. The SFI, for example, requires all of its members to subscribe to a set of principles and guidelines; however, there is no external monitoring of company performance. The SFI principles and guidelines are a mixture of general concepts and some tractable measures of company performance in meeting broad SFI objectives. They involve on-the-ground practices, as well as management systems, and are forestry industry-specific, thus differentiating them from ISO-style generic standards. The system developed in Canada by the Canadian Standards Association and the Standards Commission of Canada is also of this type.

SUSTAINABILITY REQUIREMENTS FOR THIRD-PARTY CERTIFICATION AND MONITORING

Many groups advocate the use of ecolabeling and certification of on-the-ground forest practices by third parties to ensure forest sustainability.[1] Although the certification program of the Forest Stewardship Council (FSC) has received wide attention, this approach is not new. The American Tree Farmers (ATF) have had a third-party certification of forests and forest practices for many years. Currently, there are about 95 million acres (almost 40 million hectares) of ATF-certified forest in the United States. The FSC approach is to develop global forest practices and principles and then to endorse national or regional standards developed by FSC-accredited certifiers that monitor on-the-ground performance. Under this approach, the third-party, FSC-accredited certifiers assess whether a company's forest management activities conform to the standards established by the certifying organization. The FSC currently identifies a number of "auditing" firms that are FSC-qualified to undertake forest certifications. At present, a small number of these firms have received FSC approval; for example, Scientific Certification Systems, the Soil Association, and the Rainforest Alliance (specifically, its Smart Wood activity).

Forest certification involves an examination of management systems and policies in place, in much the same way as the ISO process, but it also evaluates on-the-ground procedures and practices as well as issues such as worker safety and social issues such as local community involvement. The certification process, however, typically has a second component, that of certifying that products are made from wood from sustainably managed forests.

Strictly speaking, product certification is not a strict requirement of forest certification since forests can be managed sustainability without product certification. However, the process of ecolabeling involves tracing the raw material from the certified forest to its manufacture (maintaining a chain-of-custody) as a product in order to verify that the products come from sustainably managed forests (Bourke 1996).[2] Since it is often maintained that consumers will be willing to pay a premium for products made from certified wood (ecolabeled products),[3] certifying that the material was produced by a sustainably managed forest is a prerequisite to capturing financial returns from sustainable management.

Establishing a "chain-of-custody" for products adds the extra dimension of following an individual log to the mill and through the production process so that the processed wood product can be identified as "produced from a sustainable forest." This step is often quite complicated and costly in practice. For example, if the logs are processed in a mill that also processes wood from noncertified operations, the products from the certified and noncertified wood must be kept separate. This can entail running a separate "line" within the mill or, at a minimum, running the certified wood in such a fashion as to allow keeping the certified wood identifiable.

Obviously, maintaining the chain-of-custody from the forest through the milling operation can be a very complicated and costly operation, especially if the mill draws its resource from a large number of land holdings and many owners choose not to participate in the voluntary certification program. In fact, strict adherence to FSC guidelines on chain-of-custody and plantation management would be extremely difficult for many companies and landowners, especially smaller ones, where forest ownership is fragmented, as in the United States and elsewhere. In the southern United States, a major mill may get only a fraction of its wood feedstock from its own forestlands.[4] The remaining wood typically comes from a host of different nonindustrial forest owners, many of whom might choose not to participate in voluntary certification. These forest ownerships also vary in that some are dedicated to wood production, while others only occasionally harvest industrial wood. Additionally, the nonindustrial owners may, at various times, provide wood to a number of different pulpmills. In this type of situation it could be very difficult to separate certified from noncertified wood once it has reached the mill and is processed into pulp or paper. By contrast, the operation of a pulpmill, for example, involves the input of millions of tons of wood fiber. In some regions, such as northern Canada, all of the wood might be provided by a single concession. If the concession were "certified," then all of the pulp and paper coming from that mill could be labeled as coming from a certified forest.

In Sweden, the large companies have agreed to FSC certification, while the small private landowners' association has decided not to participate. To address this type of problem the FSC has agreed to certify products if some threshold proportion of fiber— for example, 65%—can be identified as from certified forests. Alternatively, certain Swedish mills might restrict their use of wood solely to that produced by certified forests. However, this might not always be feasible. An interesting case is the situation faced in some of the Nordic countries that are now receiving in excess of 20% of their industrial wood supply through imports from the Baltic countries and Russia. In this case, even if all of the forests of a Nordic country were certified, there would still be the problem of separating the imported, noncertified wood from the certified wood. Again, this results in a product—pulp—that either cannot be certified or receives only a qualified certification such as "produced from 80% certified wood."

Thus, the certification of individual forest sustainability can carry over into the "labeling" of a product as being made from "certified" wood only if a "chain-of-custody" is maintained; maintaining the chain-of-custody is often costly and in some cases may be infeasible.

There is also little evidence at this juncture that consumers are discriminating against timber products based upon the origin of their wood fiber. A few retailers have sought to carry certified wood products but have not found a groundswell of consumer interest. Home Depot in the United States recently abandoned, at least temporarily, its program for carrying certified wood products. Retailers in the United Kingdom have found it difficult to stock their inventories with certified wood products. Based on the experiences of these firms, quality and price would appear to be the principal drivers in consumer choices (Sutton 1996). However, some major U.K. retailers appear committed to attempting to stock only certified wood. At this time the extent of a market for certified wood and the willingness of consumers to pay a premium for that wood remains unresolved.

There are questions, too, related to the use of certification as a barrier to international trade. Some believe that certification systems will impose a nontariff barrier to trade by using certification as a justification to restrict trade (Bourke 1996). Although some European countries have shown interest in considering import restrictions on wood products that are not from sustainably managed forests, no formal restrictions have actually been put in place (Gluck, Byron, and Tikkanen 1997). Also, there is some question as to how such restrictions might be treated by the World Trade Organization (see Elliott and Donovan 1997).

To date, only 25 million hectares of forests have been FSC-certified. These forests produce approximately 3.5 million cubic meters of wood per year, a small volume compared to the 1.7 billion cubic meters of industrial roundwood produced every year (Ghazali and Simula 1996).

ENDNOTES

1. The terminology is changing. The FSC, for example, is certifying forests as "well managed" rather than "sustainability managed."

2. Although ecolabeling is an activity associated with products, and not the forest directly, the need for chain-of-custody provisions in forest certification appears to relate only to the provision of an opportunity for ecolabeling of products made from wood produced in certified forests. Chain-of-custody provisions in themselves appear to have nothing to do with improving the direct management of the forest. Ecolabeling offers the forest owner the possibility of recapturing some of the extra costs associated with forest certification.

3. A German study assumed a premium of 5%, while a U.K. study assumed 13% (Bourke 1996). Also see Ozanne and Vorsky (1996).

4. Although the SFI is moving toward requiring loggers to conform to sustainable forestry practices, logging is only a part of the menu of best practices.

4

Cost Implications of Sustainability Practices

The cost implications of systems with objectives ranging from self-monitoring sustainability to certification warrant further discussion. Incremental costs are to be found in the following areas, each of which is described in this chapter.

- Costs of introducing the new system
- Costs of implementation, including on-the-ground costs
- Costs of third-party certification and monitoring
- Costs of implementing the chain-of-custody
- Changes in mill costs associated with processing and segregating products, while maintaining the chain-of-custody control

COSTS OF INTRODUCING THE NEW SYSTEM

For any of the systems aimed at improving forest sustainability, there will be costs associated with introducing a new system. The criteria and indicator approach, for example, would require that countries put into place a system designed to monitor their forest and its condition. Such a system is generally in place in most developed countries but not in place in many developing countries. Moreover, for many of the indicators identified by the Santiago Declaration, suitable data are not available.

With respect to other systems, the use of the ISO-approved systems would probably be relatively easy and cost little for a modern firm to put into place. A third-party firm could then be used to certify ISO compliance. Additional costs to the firm would result if any gaps or omissions in the firm's existing operations needed to be rectified. For a more focused system, such as the forest-specific system advocated by the Canadians, the costs would depend in part on the complexity of the actual system developed. Voluntary systems, such as the Sustainable Forestry Initiative in the United States, are relatively inexpensive to put in place. However, expanding landowner assistance and logger education programs will require additional corporate funding, and the actual on-the-ground implementation of new practices will undoubtedly increase costs substantially in many instances.

COSTS OF IMPLEMENTATION, INCLUDING ON-THE-GROUND COSTS

Costs of implementation will depend on changes dictated by the new sustainable management plan compared with earlier procedures. (Table 2 in Chapter 6 provides some estimates of the relative costs from the implementation of sustainable management in various regions.) One study of a number of alternative forests in the Nordic countries found that the introduction of a sustainable management and harvesting implementation system increased management costs by between 5 and 40%. A major component of the increased cost resulted on lands where harvesting was eliminated or severely reduced, thereby reducing the harvest volumes and raising the average cost of a unit of wood harvested.

COSTS OF THIRD-PARTY CERTIFICATION AND MONITORING

In many of the proposed sustainable forestry systems, monitoring is internal and embodied in new procedures put in place within the firm or management unit. However, in the case of the third-party certification approaches, the costs of receiving independent certification may be substantial and are likely to involve some degree of ongoing monitoring along with its associated costs. Typically, certification involves an initial, thorough evaluation of the operation, followed by occasional on-site inspections after forest practices have been certified.

Significant economies of scale appear to exist in the certification and monitoring process. This could put small firms at a substantial competitive disadvantage and might spur mergers, acquisitions, or cooperatives to capture the scale economies created by the certification process. Hence, the costs for a small firm are likely to be proportionately greater than for a large firm.

Although the experience on actual certification is limited, some examples provide the basis for preliminary estimates of the cost of an assessment. According to Upton and Bass (1996) the minimum assessment cost is about $500. For natural forests a $0.40 per hectare initial assessment cost would be expected plus about $0.15 per hectare for subsequent visits. How frequently subsequent visits would be is left unstated but presumably depends to some extent on the situation.

In an example of the assessment costs given for a 100,000-hectare natural forest concession in the tropics, the total assessment cost was $130,000 over a six-year life, or $22,000 per year for each of six years (Upton and Bass 1996, 105). Estimating that the annual harvest is about 125,000 cubic meters, the cost per cubic meter harvested would be only a modest $0.18. (This calculation assumes a concession system such as exists in the Asia-Pacific region with a potential harvest of 50 cubic meters per hectare and a harvesting cycle of forty years. However, if the situation were more similar to that of the Neo-tropics, the average harvest might be nearer to 8 cubic meters per hectare on a sixty-year cycle. Under those conditions the cost per cubic meter would rise to about $1.70 per cubic meter. Although almost an order of magnitude higher, the costs are still relatively modest; for example, the common royalty payments in the Asia-Pacific are about $10 per cubic meter.)

In another example, the cost of certifying the seventy-seven-hectare Dartington forest in the United Kingdom, was $1,000 or about $15 per hectare (Upton and Bass 1996, 112), assuming that sustainable harvests of 5 cubic meters per hectare generate an annual harvest of about 245 cubic meters or $4 per cubic meter. However, if the subsequent annual cost drops to one-fourth of that, the cost would be about $1 per cubic meter harvested.

COSTS OF IMPLEMENTING THE CHAIN-OF-CUSTODY

The costs of implementing a chain-of-custody procedure are likely to be a function of procurement practices and land ownership patterns. Where the operation involves harvests from sustainable forests going to a single mill for processing, the costs might be modest if the mill can run these materials through in a single lot without disrupting their overall scheduling. The milled product can then be transported to the final user—for example, a retailer marketing lumber from sustainable forests—or for use as input into a product that is certified or ecolabeled, such as a door.

With more complicated logistics, the costs of properly implementing the chain-of-custody may become prohibitive. For example, the cost of maintaining a credible chain-of-custody could become prohibitive for a large timber company that has wood from various sources going into a number of different mills producing a number of different final products.

These types of problems would probably persist for commodity products including much of dimensioned lumber and woodpulp. Earlier we noted the difficulty faced by Nordic countries where a substantial portion of the wood feedstock is imported from countries that are unlikely to subscribe to any formal certifiable forestry program in the foreseeable future.

CHAIN-OF-CUSTODY CONTROL AND CHANGES IN MILL COSTS

A problem related to the chain-of-custody costs is the cost associated with changes required in the mill operation to segregate certified wood lots from the normal flow of noncertified wood. This issue was addressed tangentially above. In some cases, the costs of treating the certified wood separately would be prohibitive, as in a pulp mill where it is not feasible to run the process for a time with certified wood that could then be associated with a separate unique output. In other cases, the mill might readily adapt by creating a separate line for certified wood, where the volumes are appropriate, or by making separate runs that are consistent with the overall flow of work within the mill. However, there are additional costs associated with maintaining separate inventories and separate bookkeeping operations.

5

Some Consequences of Forest Sustainability Issues

Progress toward sustainable forestry is not uniform, either within individual countries or across the global temperate forest area. Some countries already have in place a set of laws and regulations that serve as the backbone for sustainability while others are just now considering how to make the necessary changes to address sustainability. Additionally, sustainability increases the costs of managing, harvesting, and processing forest products. Countries that already have, or are soon to have, their sustainable forestry systems in place will experience changes in production, costs, and competitiveness as a result. These and other consequences of the current state of sustainable forestry are discussed in this chapter, which builds upon the discussion of cost implications in Chapter 4.

CHANGES IN PRODUCTION

Production shifts associated with moving toward a unified concept of sustainable forestry can be manifested in at least two ways. First, production can shift between regions of an individual country, affecting regional competitiveness and production internally. Second, production changes at the national level can occur, reducing the competitiveness of one country while increasing another's.

For example, some in the United States have argued that the current set of environmental laws—including the Endangered Species Act (ESA) of 1973—make forestry operations in the United States among the most regulated in the world. The ESA was passed with the intention of preventing the extinction of animal and plant life in the United States. The act applies to all lands, federal and private. In 1989, the northern spotted owl, a species native to the forests of the Pacific Northwest, was listed as threatened by the United States Fish and Wildlife Service. One result of the listing has been the reduction in softwood timber removals in the Pacific Northwest, shifting production to the U.S. South (where other species raise concerns) and Canada.

The most recent U.S. Forest Service Resource Planning Assessment Projections for timber removals indicate a decrease in softwood harvest in the Pacific Northwest of almost 28 million cubic meters between 1991 and 2000. Nearly all of this reduction is due to spotted owl protection. Despite the change in management and subsequent decrease in timber removals caused by the ESA, demand for wood products in the United States and elsewhere continues to increase. Accordingly, other regions are meeting that demand. Softwood removals in the U.S. South and North have increased dramatically, as have imports from Canada, while U.S. hardwood removals have increased as well. Additionally, large areas of western forest (especially public lands in the Pacific Northwest) have been almost totally removed from timber production, providing incentives for more intensive forestry in the South.

Along with shifts in regional production and supply, changes in price and competitiveness can be linked to the reduced softwood harvest in the Pacific Northwest. Stumpage prices for softwoods and hardwoods have increased in the South and North, increasing costs for processors in these regions. However, high wood prices provide increased revenues for forestland owners and incentives for increased regeneration and silviculture. Industry has adjusted by building new processing facilities and/or increasing landholdings in the South. As a result, more harvests are occurring on nonindustrial lands, affecting the competitiveness within regions and between industry and private owners.

Despite the increased harvests in other regions of the United States, offsetting some of decrease in the Pacific Northwest, there is projected to be a net reduction in the softwood harvest for the country overall between 1991 and 2000. Most of this difference will be made up by an expected increase of 14 million cubic meters of industrial wood imported from Canada. Yet Canada also is projected to experience decreased harvest levels in British Columbia, due to the Forest Practices Code of 1996. The Resource Planning Assessment Projections note that reductions in the Canadian softwood harvest will likely fall in the short run, but then rebound and increase over time. If Canadian harvests decline, either in the short term or long term, increased U.S. imports from Canada will come at the expense of reduced Canadian exports to third country markets. Thus, reductions in harvest in one region of the United States, caused largely by efforts to protect biological diversity and increase sustainability, may have major impacts not only within other regions of the United States and Canada, but outside North America as well.

Just as with broadly defined sustainable forestry practices, an unintended consequence of uneven implementation of specific policies such as certification may be a reduction in timber availability in one region or country, to which other regions could respond by increasing their harvests (at least in the short run). As noted, reduced harvests in the United States have generated increased timber harvests in Canada. Similarly, increased harvests in eastern European countries are resulting in reduced harvests in some western European countries. Both short- and long-run differences will affect domestic and international production, prices, and competitiveness, as well as broader issues of cost and international trade consequences.

COMPETITIVENESS CONSEQUENCES

Changes in the way sustainability is defined and regulated are likely to have differential effects on the costs of various producers, with inevitable changes in relative costs. The costs of certification can comprise a host of factors, including not only forest practices but also costs related to administration—such as the costs of the certification audit (which may be greater with many small ownerships), the costs of the chain-of-custody procedures, and so forth. Additionally, depending on how sustainability is applied to old growth and plantations, these regions may be advantaged or disadvantaged. Thus, the costs are closely related to the way that the problem is framed and to the "rules of the game."

In general, other things being equal, the types of sustainability criteria that are emerging appear to make it easier and less costly (per unit of output) for large ownerships to adapt than for small ownerships, and less costly for government-owned forests to adapt than for privately owned forests. For example, it would almost certainly be easier for Canada, with the vast majority of its lands in public ownership, to implement uniform sustainability standards and chain-of-custody tracking than for the United States, with its millions of forest owners. The chain-of-custody issue is likely to be difficult and costly where small forest ownerships prevail and if many within an area choose not to undertake the new practices. This situation could occur in the U.S. South.

Countries like Finland would probably find it less costly than the United States to implement "traceable" sustainable forestry standards despite their large number of small ownerships because most of the small holdings tend to belong to well-developed cooperatives. Finnish cooperatives have a history of setting standards and negotiating prices with the major buyers. However, the Nordics may face difficulties brought on by the mixing of domestic wood with imports from Russia, Estonia, Lithuania, and others—about 20% of their total supply. Such mixing occurs at many of the pulpmills in the Nordic countries. Countries like Chile and New Zealand probably have some advantages in instituting sustainability standards due to the homogeneity of their management operations—for example, exotic plantations—and a relatively limited number of commercial ownerships. This advantage will depend upon plantation forests; being treated as sustainable and not overburdened with restrictions on herbicides and pesticides.

Although Germany and France may also experience some of the disadvantages of diverse private ownership, many of the sustainability standards represented in the proposed certification guidelines of the Forest Stewardship Council (FSC) are more consistent with the uneven-aged management common in European forests than with the even-aged management practiced in North American forestry. For example, clearcuts, which are often treated unfavorably by "sustainability" criteria, are a reasonable way to harvest even-aged shade-intolerant pine forests in the United States, which require large clearings for regeneration. However, clearcutting is not as suitable a harvesting method for uneven-aged shade-tolerant spruce stands, common in Europe. If sustainable criteria are biased toward uneven-aged management against even-aged management systems, as appears to be generally the case, European forestry is likely to be competitively advantaged at the expense of regions with forests that lend themselves to even-aged management, as in much of North America.

Overall, the forest industry in most countries appears to have both comparative advantages and disadvantages in meeting various sustainable criteria based on its ownership pattern and the inherent aspects of its forests. Most temperate countries are modifying forest management approaches to adjust to the new policies that are being instituted domestically, and often exceed governmental standards to meet some broad industry guidelines. For example, harvests are being modified in the Nordic countries to reduce damage to certain habitats with rich or unique biodiversity, such as wetlands. Foresters are using "green tree retention" and increasing reliance on natural regeneration, rather than planting. However, changing economic considerations are also driving many of these changes. In the Nordic countries, lower current and anticipated mill prices, brought about by raw wood imports from the former Soviet Union regions, have resulted in relatively weak wood prices in these countries, thus providing less incentive for costly investment in future wood supplies.

Canada also seems to have both advantages and disadvantages. A disadvantage is found in the inclination to view harvests of old growth forests as unsustainable forestry. An inherent advantage is because of Crown ownership of most of its forests. Uniform owner-ship permits rapid changes, as can be seen in British Columbia where forest practices are changing substantially in response to the new Forest Practices Code there. The intensity of harvesting is being reduced in some areas, as approval of harvests on one site depends upon the conditions in adjoining sites. For example, har-vesting might not be allowed on one stand until adjoining sites have achieved a target level of regeneration, such as a twenty-year-old stand. Less intensity in one area, however, is likely to deflect harvests to less accessible areas and necessitate additional road building and harvesting in other areas. The new regulations in British Columbia appear to favor less intensive harvesting on a larger portion of the forest land base.

INTERNATIONAL TRADE CONSEQUENCES

Costs associated with sustainable forestry are likely to vary system-atically by country because regions tend to face different policy conditions and somewhat different physical and natural condi-tions, and also because the costs of achieving sustainability are not likely to increase proportionally everywhere. In the face of a sus-tainably managed forestry standard, firms (and countries) will be advantaged in international trade if their costs increase less than those of their competitors and disadvantaged if the reverse is true. This advantage, as discussed above, will differ depending upon the nature of the criteria required for sustainability or for certification. It is surely possible that a number of different standards of sustain-ability may exist at one time. In fact, this is the current situation. For example, most large U.S. producers are now committed to the Sus-tainable Forestry Initiative (SFI), and elements of the U.S. industry are working toward the development of a generic ISO 14000 stan-dard. However, there is little support for third party certification of the type offered by the Forest Stewardship Council (FSC). To a large extent, the market may ultimately determine the choice of criteria. If markets prefer one standard, this standard will tend to dominate.

Given a certification standard, firms and countries with low costs of certifiable production would likely have a comparative advantage in this market and would be expected to focus their production on certified wood. Those with high additional costs associated with certification would find themselves at a comparative disadvantage in international trade and would be expected to resist moving to certification practices. These latter countries might choose to specialize in producing wood for the nonlabeled market. The producers with and without the comparative advantages could be systematically bunched in different countries. For example, if old growth is more costly to certify than second growth, then the United States and the Nordic countries might have a certification advantage compared with Canada and Russia. Or if even-aged management is treated as a nonsustainable practice, then European countries with species amenable to uneven-aged forestry might have advantages in certain markets over certain production regions of the United States. Thus, with the advent of markets for labeled products as well as noncertified products, firms, and perhaps countries, can be expected to specialize in one product or the other.

In fact, countries may have advantages in some aspects of sustainability and disadvantages in other aspects. The effects on international trade will depend upon a combination of cost increases and market acceptability that is likely to emerge from the introduction of various changes. These various facets have been discussed in greater detail elsewhere in this book.

Finally, if certification is to be voluntary, producers must meet the standards established by domestic certifiers in their respective countries, but not all will choose to meet the still higher standards that may be associated with international third-party certification. Thus, costs will rise more for some producers than for others, but , on average, overall costs will rise. For an unchanged demand, this suggests a higher wood price and decreased global consumption of wood as consumers turn to wood substitutes.

Certification as a Trade Barrier

Some of the features of plantation certification appear to move in the direction of reducing the intensification of and therefore the productivity and effectiveness of forest plantations in their role of removing logging pressures from native forests. FSC certification, for example, appears biased toward promoting natural and native forests. Plantation forestry is accepted in large part because of its ability to take pressure for industrial wood production off native forests. However, an inherent tension exists because plantations are most effective in this role when they are intensively managed and very productive—that is, when forest plantations are more like agricultural cropping than like a natural forest. Thus, it is forest plantations as agriculture that offer the greater promise to reduce pressure on native and natural forests (see Sedjo and Botkin 1997). However, intensively managed forests, utilizing fertilization, genetically improved stock, and biotechnically improved pesticides are objectionable to FSC forest guidelines.

Furthermore, a decision to systematically resist intensive management has important regional implications. The potential for high yield from plantations varies greatly across forests and regions. Many boreal forests, for example, have only limited potential as intensively managed systems due to climate and soil conditions. In other regions, particularly in the tropics and subtropics, the potential productivity gains from intensively managed forests may be huge. Certification that rejects intensive management—with genetically improved seedlings and the application of biotechnology to production—is bestowing a comparative timber production advantage on countries with low forest productivity potential, while leaving at a competitive disadvantage the countries blessed with a high productivity potential. Such restrictions will redound upon the trade potential and patterns of these countries.

Industry Adjustments to Changes

Even as the forest products sector is adjusting to a changing international and domestic institutional environment, considerations of sustainability and sustainability concepts are changing the "rules of the game." The forest products industry faces increasing costs related to new concepts of sustainable management, monitoring, and maintaining a chain-of-custody, as well as from new market pressures for delivering products from sustainable forestry sources, potentially differential prices based on source of supply, and shifting competitiveness in international trade. Furthermore, these changes are continually occurring, and companies are struggling to determine how best to adapt to their changing industry environment. Producers are currently trying to assess how a particular change is likely to affect their competitive position, as well as how and under what conditions they might be in a position to influence the rules of the game. In sum, producers are struggling to position themselves so as to be a beneficiary of the changes rather than a casualty.

MARKET DEMAND

In addition to changing domestic policies, the marketplace creates additional influences to which the forest industry must adjust. The media attention given to international concerns about forest sustainability has helped to legitimize environmental claims among consumers. A great deal of the willingness of firms to change land management procedures appears to be driven by concerns about the acceptability of their product today and in the future in foreign markets, especially some European markets that are expected to become increasingly "green" in the future. The consumer issue, however, is not limited to Europe. The Canadians in particular appear to be very sensitive to this issue.

The total global demand for industrial wood today is in excess of 1.5 billion cubic meters annually. Of this, it is estimated that only about 3.5 to 4.0 million cubic meters, or a fraction less than 1% of the world's production, is third-party certified (NFC 1996). The direct market for ecolabeled products appears to be modest but potentially important in some regions, such as the United Kingdom or Germany. Thus, the current market demand for "certified wood" products is quite limited, consisting mostly of niche markets, largely in Europe, and probably represents a very small fraction of total demand—no more than 1 or 2% of the industrial wood market.

However, certain consumers of "commodity" products have also raised sustainability concerns. For example, the U.S. newspaper industry is very sensitive to the forest sustainability issue. Not only is it beginning to face competition from the electronic media, but also some of its customers are concerned about the use of trees in newsprint. The large purchasers of newsprint are raising this concern with the forest industry. Some users of newsprint have indicated a desire to use fiber that they can claim as coming from sustainably managed forests. Thus, the market for sustainably produced wood and wood fiber may be larger than simply the specialty niche markets where it is often noted.

In the future, most of the growth in aggregate demand for industrial wood is likely to come from the developing world, where the concerns about certification and other sustainability measures are relatively small. Thus, there will almost certainly be a very large market for noncertified wood into the foreseeable future (Bourke 1996), especially if noncertified wood is cheaper due to a price premium associated with certifying wood. Under these circumstances it would not be surprising if much of the world's wood continued to be produced by firms practicing "traditional" forestry—that is, forestry that does not incur the higher costs associated with certification.

PRICE PREMIUMS FROM CERTIFICATION

A number of studies have been undertaken to determine the extent (if any) to which consumers are willing to pay a price premium for certified wood. Price premiums are most likely to exist in small specialty markets for products consumed in countries with strong "green" preferences. These markets are apt to tie product quality differences to a certification label. Such products are likely to be produced by firms that have a comparative advantage in producing wood that can easily be certified and moved into the market. An early example is the Seven Islands company of New England, which is producing only certified wood. Additionally, if many firms produce certified wood as a promotional tool, production of sustainable wood could easily swamp the modest niche markets for certified wood. However, where there are costs to meeting this standard and no price premium exists, some firms, or countries, will almost certainly choose to focus production on the demand of that large segment of the market that is indifferent to sustainable management. It is easy to envisage pulp producers from the Asia-Pacific region moving into this market. Furthermore, it is conceivable even in the United States that in the event of onerous chain-of-custody requirements, some mills will specialize in producing from sustainable forests while others will produce from the harvests of lower cost, traditionally managed forests.

Economic analysis indicates that price premiums may not exist in some markets even though some consumers are willing to pay a premium. A price premium depends not only upon the "demand side"—that is, the willingness-to-pay of some consumers—but also upon the "supply side"—that is, the portion of the total market willing to pay the premium compared with the portion of total production that is produced using sustainable management. Given a situation where a large portion of production is likely to fall under a certification initiative but where the majority of consumers are unwilling to pay a premium for certified wood (as perhaps is the case in the United States, and almost surely in third world countries) the markets for major products and commodities probably cannot maintain a price premium. Surplus commodity production will drive the price down, eliminating any premium, until consumers indifferent to sustainable management consume the surpluses. However, for certain sensitive or specialized products, a price premium may well exist. A willingness to pay a premium might be found, for example, for specialty items and perhaps even for newsprint, where the demand for the sustainable product dominates the specialty market. Also, for products being exported to important green markets, the premium is more likely to exist due to the dominance of demand for the sustainably produced product.

In light of the uncertainty and changes faced by firms and countries that must grapple with concepts of sustainability in forest management, the use of specific criteria to precisely determine the components of sustainable forestry becomes critical. For example, if harvests from old growth were viewed as violating sustainability, countries such as Canada and Russia would be seriously disadvantaged. If uses of herbicides and pesticides were restricted, the competitive position of countries with plantations would be seriously undermined; countries such as Chile, New Zealand, and the U.S. South would be competitively disadvantaged. Similarly, if chain-of-custody procedures are required, countries with many owners and mills, such as the United States, Finland, and France could be seriously disadvantaged since the implementation of satisfactory chain-of-custody procedures is likely to be costly under this type of land ownership structure. Thus the consequences of sustainable forestry depend very importantly on what specific criteria are chosen as the "rules of the game." And these criteria are, at least to some extent, arbitrary or determined by political considerations.

6

Domestic and International Policy Responses to Sustainability in Selected Temperate Countries

Aside from the international dialogue on forest sustainability, there have been a host of changes in forest policy and in land management decisions within temperate forested countries. Many of these changes have occurred within the past six years since UNCED, while others predate that event. Nevertheless, the themes invoked in international protocols are having an effect on domestic forest policies and thus forestland management, as well as on the behavior of private parties and nongovernmental organizations (NGOs). To some degree, governments have responded to internationally articulated concerns and understandings with changes in domestic forest policies. Additionally, the international governmental activities provide credibility and standing for the sustainable forestry concerns and activities of private parties and NGOs, including the forest industry, which has in many countries initiated sustainable forestry activities.

This chapter describes some of the major impacts of various sustainable and certification forestry programs on the eight countries studied: changes in domestic forest policies; changes due to biodiversity and environmental standards; and a summary of the relative impacts on practices, land ownership patterns, and forest conditions. Appendix A details the rest of the findings of this study about these eight countries.

CHANGES IN DOMESTIC FOREST POLICIES

Most of the countries examined have recently instituted new forest policy or important sets of regulations, or they were in the process of doing so. (The countries selected for closer examination were chosen on the basis that they are major temperate region producers and traders of industrial wood.) For example, Finland and Sweden recently modified their forest policies to give the protection of biodiversity equal status with the production of industrial wood. This is being reflected on the ground by some modification in forestry practices and a change in the designation of protected areas, with both countries placing a greater part of their forest lands in protected area status. Some of these protected areas are wetlands that might have been drained and harvested under earlier policies. Other areas being protected include older forests and forests of mixed species. In addition, planting and harvesting practices are being modified. Harvesting practices are becoming more environmentally benign, with fragile areas either not harvested or experiencing a modified harvest. Harvests are leaving more standing trees, particularly nonconifers and dead trees and snags (for habitat purposes). Natural regeneration is replacing planting in many places and mixed-species stands are being promoted. While some of these changes are being driven partly by financial considerations (a view offered by the Nordic countries[1]), environmental and biodiversity considerations are also having an influence.

As noted above, in parts of Europe, some argue that serious changes in forest practices are not warranted because a sustainable type of forestry is being or has been practiced for centuries.[2] However, this has occurred in a context with little or no remnant old-growth forest, and changes are under way despite this opposition. Anticipated new legislation in France, for example, is expected to upgrade the role of forests in maintaining biodiversity and wildlife habitat. Previously, the primary role of forests under the law was as a source for wood.

Also, the Canadian Forest Accord (1992) provided for the creation of a multiple stakeholder National Strategy Coalition that commits the signatories to sustainable development of Canada's forests. All provincial and territorial governments, which are responsible for managing the vast majority of Canadian forests, are signatories to the accord. British Columbia , in Canada, has recently instituted major changes in forest policies. A new 1996 B.C. Forest Practices Code covers all public forests in British Columbia.

The United States is one country that has not made major *de jure* forest policy reforms at the national level. However, the actual management of U.S. public lands changed dramatically in the first part of the 1990s due to court interpretation of existing legislation, particularly the Endangered Species Act.[3] State and local regulation of forest practices prevails in most areas of the United States, and although state forestry codes have become more restrictive, they have not resulted in management changes of the magnitude seen on federal lands.

Changes in policy are also evident in countries where plantations figure prominently. Although some early sustainability criteria precluded plantation forests, the trend is to discourage native forest conversion to plantations but to recognize that plantations can be part of a national strategy for sustainability.[4] In Chile, the policy environment affecting forestry has undergone change such that harvesting permits are required and afforestation is necessary to qualify for preferential tax treatment. Also, New Zealand enacted a Resource Management Act in 1991, which is essentially a national land use planning law designed to be "effects based," that is, designed to mitigate against specific environmental impacts. In 1993, New Zealand tightened regulation of activities on native forests. Native forests are now almost entirely protected in New Zealand, and are likely to be tightly regulated in Chile pending the outcome of proposed legislation this year.

In addition to governmental policy, the forest industry in many countries has taken the initiative to develop national or regional sustainable forestry plans or initiatives. This action is driven in part by the industry's recognition of new social concerns over forest management and in part by concerns over the ability of a region to maintain its position in domestic and foreign markets, especially "green" foreign markets. The changes are also being driven by concerns over the potential role of environmental NGO certifiers, such as the Forest Stewardship Council (FSC). Industry would like to spearhead its own sustainable forestry systems.

Industry activities have already had an effect on land management decisions and outputs. Both Sweden[5] and Finland have been working to develop domestic certification systems and are also working with Norway to create a Nordic Certification Organization, which would provide coordination and communication among the Nordic countries on certification issues. Also, Canada's industry, working through the Canadian Standards Association, has developed a Sustainable Forest Management standard that covers the Canadian forest industry. Most of the Canadian forests are public "Crown" lands so standards can be more easily applied. A parallel activity being led by the Canadian Pulp and Paper Association is the idea of developing a "sustainable paper industry."

In the United States, major portions of the forest industry have voluntarily agreed to adhere to the Sustainable Forestry Initiative, sponsored by the American Forest and Paper Association (AF&PA). The industry initiative was originally driven by the large companies that both produce and market. Many individual firms are undertaking related types of activities that are consistent with the AF&PA initiative but that have some unique aspects.

For example, Champion International holds an annual forum to discuss the activities and accomplishments of its corporate sustainable forestry activities. The Champion program, unlike the ISO programs, directly addresses forest management activities as well as generic management systems. Additionally, Champion does use external people as a quality control to provide an external opinion of their on-the-ground practices. However, the use of third parties is strictly to assist Champion in assessing its own program and is not part of an external certification program.

A comparison of the initiatives in the various countries is difficult. For example, the initial conditions are often very different. Some countries, such as the United States and Canada, had set aside large forest areas well before the 1992 UNCED meeting. Others, such as the Nordic countries, had set aside only small areas on which harvests were restricted. Additionally, nonforest policies can become important in restricting harvests and promoting other forest values. In the United States, for example, the Endangered Species Act, which was not designed to restrict or regulate harvests, has resulted in the de facto set-aside of additional millions of acres of old-growth forest in the Pacific Northwest.

IMPLICATIONS FOR LAND MANAGEMENT DECISIONS

Most of the proposed changes directed toward forest sustainability imply increasing costs for landowners and producers. Often, on-the-ground costs are increased due to biodiversity considerations, which involve the setting aside of large areas of forest for riparian zones and wetlands and the retention at harvest of some "green trees." These changes lower the overall harvest volumes from a given area and therefore result in a corresponding loss of revenue. Table 1 presents some recent cost estimates in various countries.

The approaches and costs being estimated are not always consistent across studies. However, all the studies estimate costs associated with stricter, locally imposed environmental standards, except the Kajanus and Karjalainen (1996) study for Finland, which estimates the costs of meeting the third-party certification standards of the Forest Stewardship Council.[6] Firms in the United States commonly estimate the increased costs associated with the Sustainable Forestry Initiative to be on the order of 5 to 10%. In a recent paper, Greene and others (reported in Michaelis 1996) estimated the increased costs for private landowners in the Pacific Coast region of the United States that result from the Endangered Species Act and other regulations to be from 5 to 14%. Third-party certification schemes would include additional costs such as those associated

Table 1. Some Cost Estimates of Increased Environmental Standards.

Country	Study	Cost increase
United States	Greene and others (1995)	5–14% cost increase
Finland	Tikkanen (1997)	12% decrease in net income to meet new biodiversity standards
Finland	Kajanus and Karjalainen (1996)	18% decrease in net income to meet certification standards
Canada	van Kooten (1994)	$3 cost increase/cubic meter
Canada	Haley (1996)	$8 cost increase/cubic meter (about 15% cost increase) + other costs
United States	Abt and Murray (1997)	50% of the harvest can be produced at an on-the-ground cost increase of 10% or less; the remaining 50% of the harvest requires cost increases of above 10%, to a high of about 50%

with auditing and establishing chain-of-custody, as well as those due to more expensive management requirements.

In British Columbia, a host of different studies have been done to estimate the effects of the new forestry code. An early study estimated increases in harvest costs of about $220 million annually (van Kooten 1994), while a more recent assessment in British Columbia of the cost of the British Columbia Forest Practices Code (Haley 1996) estimated that the Code's implementation would increase total annual costs by about $1.5 billion dollars. In another study using a general equilibrium model, Binkley and others (1994) estimated the annual losses to provincial GNP due to the Forest Practices Code at over $1 billion annually, or 1.54% of annual GNP. Binkley (1995) estimated that the effects of the current policy regime will result in about a 23.5% long-term reduction in total harvest levels in British Columbia. Additionally, because the policy changes are focused on reducing the impacts on harvested areas, an implication of this policy is that the harvest will necessarily be spread over a larger area if harvest levels are to be maintained. Of course, diversifying the harvest will increase the incremental costs since additional investments in roads and so forth will be necessary. Whether British Columbia harvest levels, which are dictated by the provincial government, will decline as predicted remains to be seen.

In Finland and Sweden new forest policies have placed biodiversity on an equal footing with industrial wood production. Modified logging practices are reducing the area eligible for harvest and requiring that more trees be left standing on harvested sites, thereby reducing the total harvest and increasing costs per unit of production. Tikkanen (1997) estimated that achieving the biodiversity standards would reduce net *income* of forest companies by 12%. Another recent study of the Nordic countries by Kajanus and Karjalainen (1996) found that third-party certification could increase costs as much as 30 or 40% for certain land conditions. This increase would largely result from the set-asides needed to reduce harvests and thereby achieve an ecolabel certification. Also, costs would rise due to restoration targets and the alteration of certain management methods. Kajanus and Karjalainen estimated that the average net effect would be to reduce income per hectare by about 18%.

Recently, Sweden's large forest product firms agreed to accept the FSC certification standard. However, the Swedish Federation of Forest Owners, the nonindustrial private owner's group, strongly opposed this decision and is on record as rejecting the FSC standard and its certification.

In a study in progress of the softwood harvests in the U.S. South, Abt and Murray (1997) found that about 50% of the harvest could meet FSC on-the-ground certification standards at a cost increase of 10% or less. However, the remaining 50% of the harvest would require cost increases above 10%, with some approaching a 50% cost increase. These increases did not include those costs associated with chain-of-custody considerations (Apt and Murray 1997). Thus, while most studies suggest cost increases associated with new on-the-ground sustainable practices of at least 5% and often 20%, costs will clearly vary widely depending upon the conditions of the site. Additionally, none of these estimates includes the costs of maintaining "chain-of-custody" control.

To summarize, it is clear from our discussions with the forest industry and observations of their practices that, with the combination of the changes in legislation and industry practices, sustainable forestry initiatives are creating significant changes in forest management in most of these countries. Although, in some cases, the full effects of the changes cited above remain to be seen, concerns about higher costs are valid. Marginal firms could be driven out of business. Furthermore, firms share very substantial concerns about their ability to compete internationally, given the increased costs associated with changes to meet new sustainable management criteria. Higher operating costs in a region could result in investments being driven offshore. Additionally, cost impacts on the various countries will differ depending upon which particular sets of practices are viewed as sustainable or not sustainable (see Table 1). These impacts can have a large effect on the comparative competitive position in international trade of the various countries.

RELATIVE IMPACTS OF PROPOSED SUSTAINABLE FORESTRY POLICIES

As emerges from the summary of impacts in Table 2, our assessment is that the impacts of the various policies differ greatly among the eight countries studied. In general, a low impact rating implies that a policy is likely to have little disruptive effect on forest production and that the costs of accommodating this policy would be modest. A high rating implies just the opposite: the disruption of harvests would likely be large as would be the cost of implementing the policy. Here are some examples:

- The impact of limiting the harvest of old growth falls heavily on countries with large reserves, such as Canada, and is medium to low for countries that have already liquidated their natural forest reserves. The impact on the United States is rated low since most of the old-growth forest reserves have already been set aside or are otherwise protected.

- The impact of regulations on the conversion of natural forests to plantations is rated low for all countries but the United States. The United States has been converting logged-over natural, but not old-growth, forests to forest plantations, especially in the South. The other seven countries have little conversion of forest to plantation since much of their forested area is already in planted forest (that is, much of Europe). In other areas—Chile and New Zealand—most plantations are established on former crop and pasture lands.

- The restriction on the conversion of forests to other land uses is not important in any of the eight countries examined, since all have stable or expanding areas of forest.

- Restrictions on even-aged management would have large impacts on all but the European countries (that is, France and Germany) where much of the forest consists largely of uneven-aged stands with species that can regenerate in the shade of the forest floor, and where the forest lends itself to selective harvesting. The six other regions, which harvest predominantly shade-intolerant species, would experience high impacts if a restriction were placed on even-aged management.

Chain-of-custody requirements favor countries that have large forest holdings by one or a few ownerships. Limited ownerships provide centralized control whereby new sustainability "requirements" could easily be adopted throughout the forest, and thus, most or all of the wood flowing into the mill would be "certified." Noncertified wood could readily be identified in this situation and thus the costs of maintaining the "chain-of-custody" would be modest. Canada, for example, where most of the forest is owned by the "Crown," could simply impose management regulations on an entire forest, thereby certifying all wood shipped from that forest. The United States, in contrast, has decentralized ownership and control of its forests, with 59% of its forestland owned by ten million nonindustrial owners. Most mills use wood from a variety of

Table 2. Relative Impacts of Proposed Sustainable Forestry Policies on Selected Countries.

Sustainable forestry approach	Canada	Chile	France	Finland	Germany	New Zealand	Sweden	United States
Placing limits on harvests from primary or old-growth forests	High	Medium	Low	Low	Low	Low	Low	Low
Discouraging conversion of natural forests to plantations	Low	Low	Low	Low	Low	Low	Low	High
Restricting conversion of forests to other land uses	Low	Low	Low	Low	Low	Low	Low	Low
Limiting use of genetically improved stock or exotic species	Low	High	Medium	Medium	Medium	High	Medium	High
Restricting clear cutting	High	High	Low	High	Low	High	High	High
Restricting use of chemical treatments	Low	Medium	Low	Medium	Low	Medium	Medium	High
Requiring chain-of-custody tracking	Low	Medium	High	High	High	Low	Medium	High

Note: These impacts are based on existing practices, land ownership patterns, and forest conditions.

ownerships, thereby making a credible chain-of-custody that can distinguish certified wood very difficult to maintain.

ENDNOTES

1. Natural regeneration is less expensive than planting. Additionally, the market for trees previously viewed as nuisance trees—for example, birch—helps keep attention away from the species to be regenerated.

2. This view was offered largely by the French and Germans.

3. Also, the "viability" regulations used for Forest Service lands have been tightly applied in recent years and are quite stringent.   .

4. For an articulation of the argument that plantations can be used as a vehicle to protect native forests see Sedjo and Botkin (1998). The tension between the national strategy perspective and on-the-ground practices is found in the general view in the Forest Stewardship Council's International Principles and Criteria that "plantations should promote the protection, restoration and conservation of natural forests." (FSC 1997).

5. Sweden's major industrial producers have an agreement for FSC certification. However, this does not apply to the nonindustrial cooperatives in Sweden.

6. This study showed how the costs may vary greatly depending upon site and upon the amount of forest set-aside for biodiversity purposes.

7

Summary and Conclusions

In the six years since the June 1992 United Nations Conference on Environment and Development, forest sustainability has been the focus of a number of international and domestic government actions as well as private efforts. These actions include numerous dialogues, some nonbinding international agreements, a host of policy changes within countries, and the creation of several strategies to promote sustainable forestry through management standards, professional associations, and third-party certification.

It is clear that the international, national, and private approaches to sustainable forestry offer a diverse variety of perspectives on sustainable forestry. At one extreme is the view that "current practices are sustainable," based largely upon the data that show temperate forest areas increasing in many countries although forestry has been practiced (in one way or another) for nearly a thousand years. At another extreme is the view that there are many indicators that practices in these forests may not be sustainable; that careful third-party monitoring of on-the-ground forestry practices is required to ensure that sustainable forest management is practiced. Between these extremes are a number of alternative approaches. These include those represented by the multinational Montreal and Helsinki processes, which focus on national level approaches to promote sustainable forestry; the ISO Environmental Management Standard, which is focusing on developing standards for the environmental management systems within firms; industry-based sustainable management guidelines and standards such as the Sustainabale Forestry Inititiative (SFI) in the United States and those of the Canadian Standards Association; and finally an externally based set of sustainable forestry practices standards involving third-party on-the-ground certification.

One finding is that there is no clear preference or general acceptance, or even an awareness, in the marketplace among the various approaches. Nevertheless, a tremendous amount of interest and attention has been devoted to certification schemes that can be translated into ecolabeling. The sustainable forestry certification championed by the Forest Stewardship Council has been promoted heavily and has received mixed acceptance. None of the various approaches has established clear credibility, and although certain proposals (SFI, for example) have already been implemented, most are too new to have developed a clear track record. One aspect that will be a determining factor will be the cost and ease of implementation; another will be the competitive advantage created for some countries and industries by the various approaches.

Although no one approach seems to be recognized by the market as preferable, it is clear that all the the individual countries examined have responded to international and domestic concerns with changes to their domestic laws and policies to improve water quality, protect biological diversity, and implement more sensitive silvicultural treatments. Some of these policies are regulatory in nature, while others rely on tax or other incentives. In all of the temperate forested countries examined, significant revisions have been made over the past several years in the legal and/or institutional framework dealing with forest matters. In addition, significant private efforts have been initiated.

It remains an open question whether schemes involving third-party certification of forest management practices justify their costs. Although forest certification is not product certification, most types of forest certification involve a chain-of-custody that follows the wood flow to its use in a product. Unless forest certification is going to be used as the basis of product ecolabeling, the costly chain-of-custody serves no purpose as it contributes nothing to improving the condition of the forest. The labeling of products as using certified wood materials, at worst, could be highly confusing to the consuming public who may not be able to sort through multiple labels.

Despite these uncertainties, most forest stakeholders generally recognize that forest sustainability must be addressed and, furthermore, that the increasing efforts of voluntary programs in most temperate countries will continue to encourage best management practices, significant changes in public policy, and, to some degree (perhaps substantial), third-party certification.

Appendix A
Forest Policy and Regulation in Selected Temperate Forested Countries

Findings from the Study

This appendix outlines some of the findings of this study, presenting country-specific details through profiles of eight selected countries. The information was gathered by canvassing the available literature and through personal interviews with officials and representatives in the United States, Canada, Chile, France, Germany, Sweden, Finland, and New Zealand. These interviews examined how various changes in the international and domestic institutions are affecting land management decisions and how they may influence international trade patterns. Each country's profile is presented as follows:

- Overview of the country's forest situation
- Highlights of national forest policy
- Recent changes in national laws and regulations
- Discussion of sustainability guidelines, principles, and certification

Chapter 6 describes some of the other major impacts of various sustainable and certification forestry programs on the eight countries studied: changes in domestic forest policies, changes due to biodiversity and environmental standards, and a summary of the relative impacts on practices, land ownership patterns, and forest conditions.

CANADA
Overview of Forest Situation

Almost half of Canada is forested, totaling some 418 million hectares of which 222 million are classified as commercial forests. Provincial governments manage approximately 71% of Canada's forests while the federal and territorial governments manage about 23%. Thus, 94% of the forest lands in Canada are Crown lands, owned and administered by the federal, provincial, or territorial governments. The remaining 6%, which is owned by more than 425,000 private landowners, produces 19% of the annual harvest. Forest types vary from coast to coast, but spruce and fir are the predominant species types. See Table 1 for other quantitative characteristics of Canadian forestry.

Most of the private forest lands are found in eastern Canada and most have been logged-over previously. Generally, governments do not regulate timber harvests on private lands. In the west, by contrast, the vast majority of forest lands are Crown lands that have never been harvested. Since major logging activities have been underway in the west for nearly a century, the large areas of remaining old growth attest to the very large areas of forest contained in Canada. In parts of the west, logging is just beginning anew on some lands that were harvested a century ago, experienced natural regeneration, and are now suitable for harvest. An assortment of Native Indian land claims is pending in much of British Columbia and has figured into some of the conflicts over logging. Final resolution of native land claims will likely take many years.

Forest Policy

This discussion will focus on British Columbia, which produces about half of Canadian industrial wood (and accounts for 6% of the

Table 1. Summary Profile of Forestry in Canada, 1994.

Total land area	970,014,000 hectares
Total forest area	417,600,000 hectares
Forests as percentage of total land area	42.9%
Productive, closed, nonreserved forest area (PCNR)	221,560,000 hectares
PCNR as percentage of forest area	53.2%
Forest ownership:	
Public forest ownership	392,544,000 hectares (94.0%)
Forest industry	Included in Other Private (below)
Other private	25,050,000 hectares (6.0%)
Value of selected forest product exports[1]	$21.8 billion
Value of selected pulp and paper exports[2]	$12.3 billion
Value of selected solid wood product exports[3]	$9.6 billion
Forest product exports as percentage of world exports	19.9%
Forest product exports as percentage of temperate forest exports	23.9%

Notes on sources: 1. Based on 1994 FAO data for industrial roundwood, sawnwood, wood-based panels, paper and paperboard, wood pulp, and recovered paper.

2. Based on 1994 FAO data for paper and paperboard, wood pulp, and recovered paper.

3. Based on 1994 FAO data for industrial roundwood, sawnwood, and wood-based panels.

world's softwood harvest). In British Columbia, most harvesting is done under one of a number of various concession arrangements. The concession arrangements typically give much of the management control to the concessionaire, who becomes responsible for harvesting and forest regeneration. The province is divided into thirty-seven timber supply areas and thirty-four tree license areas. The Provincial Forestry Agency determines the *allowable annual cut* (AAC) for each area and for the province as a whole. Under most of these arrangements the harvest rights are related to the construction of a mill by the concessionaire and are set by the province. Usually, the province assigns de facto harvest targets to the concessionaire on the basis of mill capacity. Operators can vary the annual harvest by as much as 50% in any given year, but must remain within 10% of the prescribed AAC over a five-year period. Provincial law requires forest companies to reforest cut-over land within three to five years after harvest.

Since the 1980s, conflicts over forest management in British Columbia have become heated, particularly on the Clayquot Sound region of Vancouver Island where environmental groups have attempted to stop all logging. To address conflicts over logging and other environmental issues, a number of changes have been made in British Columbia's forestry regulations. A Commission on Resources and Environment (CORE) was established in 1992 to review and attempt to resolve land-use conflicts. The commission produced a classification system for forest lands based on its judgment of appropriate forest management intensity. At the same time, the province sought to identify areas suitable for preservation status and increased by more than half the area in preserved land status.

Recent Changes in Laws and Regulations

A new Forest Practices Code was enacted into law in 1993 to standardize forest management practices throughout British Columbia. British Columbia's Forest Practices Code (FPC) was implemented in 1996. Its objective is to limit the impact of commercial forestry, thereby maintaining a wider range of resource values. Forest Ecosystem Networks (FEN) address the landscape biodiversity objective of maintaining natural connectivity and reducing fragmentation by establishing a contiguous old-growth network and ensuring that commercial forestry does not compromise that connectivity. The networks include riparian areas, ridge tops, and connecting corridors," thereby protecting soil, water, fish, and wildlife. Visual quality objectives are also applied. The new regulations, in effect, imposed a number of new constraints on harvesting including designating the location and specifications for roads, the size of clearcuts, and connectivity requirements. According to some in the industry, the area available for commercial harvesting and the harvest-per-unit area have been reduced by 35%.

Sustainability Guidelines, Principles, and Certification

Through the Canadian Standards Association (CSA), Canada is in the process of adopting a Sustainable Forest Management (SFM) standard specific to the Canadian forest industry. The standard includes management systems but the firm can voluntarily agree to also develop on-the-ground performance standards. Monitoring of on-the-ground performance will be done by a third-party group of auditors. Acceptable performance will be "registered" but will not be termed as certified. Hence it will not be used for ecolabaling. All wood, registered or not, will be treated as sustainably managed based on the regulations of the province. This commitment comes out of the 1992 Forest Accord that committed the provincial governments, forestry industry associations, and several other interest groups to principles of sustainable forest management. Some of the action items included completing an ecological classification of forest lands, completing a network of protected areas that are representative of Canada's forests, expanding forest inventories to include a wider range of values and developing a system of national indicators to measure sustainable forest management. In 1995, the Canadian Council of Forest Ministers produced a national framework of criteria and indicators to help track the nation's progress in meeting sustainable forest management goals.

CHILE

Overview of Forest Situation

Land suitable for forests cover about 43% of Chile, but a much smaller portion would be considered productive, closed forest. The forested area of Chile is estimated at approximately 16.2 million hectares, consisting of 14.4 million hectares of native forests and 1.8 million hectares of plantations. Much of the native forest land is in protected status, with only about 7.5 million hectares potentially available for commercial use. Thus, a total of only 9.3 million hectares of forest land supports industrial forestry. Chile's forest resources are characterized by a growing inventory of plantations, of which 1.4 million hectares is radiata pine. Plantations provide nearly three-quarters of the annual industrial harvest. However, a substantial volume of firewood, harvested mostly from the native forests, is also used in Chile. Eucalyptus is the second most prominent plantation species. See Table 2 for other quantitative characteristics of Chilean forestry.

The vast majority of forest land that is commercially available is owned privately, either by corporations or nonindustrial owners. The combination of low wage rates, excellent growing conditions, a favorable tax policy, and a forest strategy favorable to plantations has attracted significant capital to Chile's forest sector during the past several years. Total harvests in Chile amounted to 17.8 million cubic meters in 1993, but production is expected to increase to 22 million cubic meters by 2010 as plantations mature.

Table 2. Summary Profile of Forestry in Chile, 1994.

Total land area	78,420,000 hectares
Total forest area	38,000,000 hectares
Forests as percentage of total land area	43.1%
Productive, closed, nonreserved forest area (PCNR)	9,300,000 hectares
PCNR as percentage of forest area	27.5%
Forest ownership:	
Public	7,980,000 hectares (85.8%)
Forest industry	810,000 hectares (8.7%)
Other private	510,000 hectares (5.5%)
Value of selected forest product exports[1]	$1.2 billion
Value of selected pulp and paper exports[2]	$0.8 billion
Value of selected solid wood product exports[3]	$0.4 billion
Forest product exports as percentage of world exports	1.1%
Forest product exports as percentage of temperate forest exports	1.3%

Notes on sources: 1. Based on 1994 FAO data for industrial roundwood, sawnwood, wood-based panels, paper and paperboard, wood pulp, and recovered paper.

2. Based on 1994 FAO data for paper and paperboard, wood pulp, and recovered paper.

3. Based on 1994 FAO data for industrial roundwood, sawnwood, and wood-based panels.

Forest Policy

A 1931 Forest Law, and regulations promulgated periodically under the Forest Law, provides the basic underpinning for Chile's forest policy. Decree 701, issued during the Pinochet regime in 1974, provided for a generous tax and incentives program for establishment of plantations. Not only are forest lands exempted from property taxes or estate taxes, but income derived from forest property receives a preferential tax treatment that extends to income derived from finished products. Decree 701 expired in 1994, was given a one-year extension, and is currently under legislative review. The Decree 701 forestry programs have increased afforestation from an average of 16,000 hectares per year to over 80,000 hectares per year. More recently, favorable market economics have stimulated further plantation establishment. New plantations are being established at an annual rate of 120,000 hectares.

Decree 701 and numerous "reglomentos" (ordinances) detail incentives and the requirements necessary to qualify for the incentives. The decree specifies the types of cutting in allowed in the twelve regions and requires a forest management plan subject to approval by CONAF (Corporacion Nacional Forestal), the Chilean forestry agency. The pending reauthorization of Decree 701 would limit eligibility for planting subsidies to small owners for afforestation projects on fragile or erodible lands. It would also increase regulation of forest management to meet environmental objectives, although some of its provisions are controversial and still being debated. For example, the pending proposal would prohibit cutting on slopes greater than 45%.

Recent Changes in Laws and Regulations

During the past five years, the policy environment affecting the forestry and environmental sectors has undergone significant change. In order to qualify for preferential tax treatment and/or reforestation subsidies, landowners must complete a forest plan. Plan requirements have been broadened to include environmental protection. Riparian buffers are typically stipulated, and harvests within one hundred meters of the headwaters of water sources are prohibited. Permits can take up to six months to obtain, however, and Chilean authorities admit that they do not have the resources to strictly enforce forest regulations in all areas.

Eligibility for cost-share programs has been tightened to include only planting on denuded forest land. Thus, only about half of the plantations currently qualify, as many are being established on agricultural land. Somewhere between 3 and 6% of plantations are being established to replace native forests. The commercial exploitation of native forests and their conversion to plantations is a controversial issue in Chile. A proposed Native Forest Law would require a landowner to post a bond of as much as $3,000 per hectare as insurance against failure to adhere to an approved forest plan. The stated purpose of its proponents is to slow the logging and conversion of native forests to other uses.

A Wildlands Protected Areas law, enacted in 1984 to regulate the use and management of then national parks, natural reserves, and biological monuments, is also in the process of being revised. A current draft revision would offer incentives to private landowners to donate their properties for preservation status.

An all-encompassing environmental law was enacted in Chile in 1994. The COREMA, as the law is known, created an agency with broad jurisdiction over environmental matters. The law was designed to prevent or reduce negative environmental impacts including destruction of natural resources. The law requires that environmental assessments for all large industrial or land development projects be drafted and approved by CONAMA, a new environmental regulatory agency. Although the law is being implemented, regulations are still being drafted and revised. Chile's Controlio, an agency somewhat similar to the Office of Management and Budget in the United States, rejected the first set of regulations drafted under the law.

Sustainability Guidelines, Principles, and Certification

Chile is a participant in the Montreal Process and is working on developing a program to put in place criteria and indicators through the Chilean Instituto Forestale (INFOR). A project to map existing resources is scheduled to be completed this year, and the results will serve as a basis for a criteria and indicators approach.

One interesting project in southern Chile stands as an example of some of the risks associated with developing new forest projects intended to meet sustainability expectations. Trillium Corporation purchased 270,000 hectares in southern Chile and Argentina, including 130,000 hectares of forest land with 30 to 35 million cubic meters of hardwoods, primarily lenge and coigue. The company's plan is to provide 200,000 cubic meters to the furniture and molding markets in Asia, North America, and Europe. The project, when fully implemented, expects to employ 800 workers and reflect an investment of $200 million. The company has developed voluminous environmental assessments and sustainable forest management plans with the vision of producing certified wood products with a green label. The company articulated a commitment to balance social, environmental, and economic aspects of the development and published stewardship principles in an entirely open and transparent manner. The management strategy calls for a 100-year harvesting cycle, no clearcuts, and no replacement of native with exotic species. They used conservative growth rates and planned to build a nursery. They convened a high ranking science committee made up of more than one hundred respected Chilean scientists, who produced seventeen studies on the project. A voluminous environmental impact assessment was produced although not required. Trillium declared 10,000 hectares would be held as permanent reserves and also calculated that 25% of the area would be set aside because of the management regimes selected for the project. No chips were to be produced as a primary product. They adopted FSC goals and were planning to obtain FSC certification. The company spent $15 million on assessments but has not been permitted to harvest any trees to date. The project was the first to test the new environmental law, and Trillium submitted plans and environmental assessments to CONAMA. COREMA approved the project but was sued by local environmental groups. Litigation ultimately led the company to Chile's Supreme Court, which struck down approval for the project.

FINLAND

Overview of Forest Situation

Finland covers about 34 million hectares, of which some 27.1 million are classified as forest. Productive forest land covers slightly over 20 million hectares. Of productive forest land, 62% is owned by small landowners or other nonindustrial owners, 8% by industry, and 30% by the public. About 15%, or 2.7 million hectares, has been set aside under protection status. During the 1980s, Finland's forest area experienced an annual gain of 5,500 hectares, as nonforest was converted to forest. Since 1950 there has been an overall increase of 22% in the growing stock and a 44% increase in total annual growth (Palo 1995). There has also been a net annual gain of 13.8 million cubic meters in timber production. Due to fires and excessive utilization, few Finnish forests are characterized as old growth or virgin. The oldest stands are likely to be second or third growth. During the postwar period, forests expanded due to peat bog drainage and reduced agricultural production. Inventories are currently at the highest levels measured during the past seventy years (the first comprehensive inventory was in 1924). See Table 3 for other quantitative characteristics of Finnish forestry.

Forest Policy

Finland's forest policy dates back to the Private Forestry Act enacted in 1928 and revised in 1967 and was originally based on the principle that forests must not be "devastated." The law obligates regeneration and harvesting in accordance with a forest management plan. The Forest Improvement Act of 1987 aimed at promoting timber production on privately owned forests. Implementation of government policy and regulations relating to private forestry is the responsibility of the Forestry Centers and District Forestry Boards under the Ministry of Agriculture and Forestry. Funded primarily by the government and supplemented by small fees from private landowners, these local authorities coordinate management activities among landowners and supervise all harvesting and regeneration on private lands. They assess applications for forest

Table 3. Summary Profile of Forestry in Finland, 1994.

Total land area	33,701,000 hectares
Total forest area	26,276,000 hectares
Forests as percentage of total land area	78.0%
Productive, closed, nonreserved forest area (PCNR)	20,032,000 hectares
PCNR as percentage of forest area	77.2%
Forest ownership:	
Public	5,929,000 hectares (29.6%)
Forest industry	1,743,000 hectares (8.7%)
Other private	12,360,000 hectares (61.7%)
Value of selected forest product exports[1]	$9.3 billion
Value of selected pulp and paper exports[2]	$7.1 billion
Value of selected solid wood product exports[3]	$ 2.2 billion
Forest product exports as percentage of world exports	8.5%
Forest product exports as percentage of temperate forest exports	10.2%

Notes on sources: 1. Based on 1994 FAO data for industrial roundwood, sawnwood, wood-based panels, paper and paperboard, wood pulp, and recovered paper.

2. Based on 1994 FAO data for paper and paperboard, wood pulp, and recovered paper.

3. Based on 1994 FAO data for industrial roundwood, sawnwood, and wood-based panels.

improvement; provide training, extension, and information services promoting private forestry; and supply professional assistance. Small fees are required of private owners.

Management of publicly owned forest is the responsibility of the Forest and Park Service, a state enterprise working under the Ministry of Environment and the Ministry of Agriculture and Forestry. A number of statutes enacted in the 1980s and early 1990s were designed to increase the area devoted to national parks and reserves and to favor biodiversity objectives in the management of public forest land.

Recent Changes in Laws and Regulations

In March of 1994 the Finnish Ministry of Agriculture and Forestry introduced a new forest policy in the form of the "Environmental

Guidelines for Forestry in Finland." These guidelines stated the government's firm commitment to sustainability; introduced multi-purpose forestry, both public and private; and placed the protection of biodiversity on an equal footing with production of industrial wood. The guidelines require changes in central elements of Finnish forestry legislation, such that management must utilize and protect publicly owned forests in a sustainable and profitable way. Under the terms of the Forest and Park Service Act (1994), that agency is explicitly required to manage publicly owned forest on a sustainable basis.

In 1996 Finland has enacted two pieces of legislation important to forestry. These are the new Forest Law and the new Environmental Law. The Forest Law applies to all public and private forests in Finland and puts biodiversity objectives on a par with industrial wood production in the context of sustainable forestry. However, this law does not appear to translate into substantial changes on the ground since the law was anticipated and was drafted to include changes in forest practices already being made by the forestry agencies. The Environmental Law is designed to harmonize Finland's environmental regulations with those of the European Union. Additionally, new endangered species regulations require small private landowners to inventory their lands to identify the plant and animal species present. In concept, if the habitat is critical to some species, the government could set it aside, although the landowner would be compensated for the taking. Finnish experts view these changes as resulting from rising global forest concerns.

Sustainability Guidelines, Principles, and Certification

Finland has committed to the Statement of Principles on Forests and Agenda 21, which came out of the Earth Summit, as well as to the resolutions of the ministerial Conference on the Protection of Forests in Europe (Helsinki Process). Finland is part of the Nordic group that has been moving toward a coordinated and harmonized Nordic system. However, each of the three Scandinavian countries has been working independently. A recent task force has worked to ready Finland to move quickly to certification if necessary. The plan is to develop a certification scheme consistent with any used by the European Union. The major impetus for certification comes from Germany, Finland's largest market, where the "Greens" are strong and are putting pressure on the publishing companies to use paper from certified forests. Finland has been using a "PlusForest" label on some products indicating that only Finnish wood is used.

The large number of private forest landowners in Finland poses a challenge for gaining acceptance of and implementing a certification program. However, the task is likely to be much easier in Finland than, for example, in the United States: Finnish owners already operate within the context of cooperatives, which have traditionally been involved in broad economic and policy issues, including forestry practices, that affect their membership. Thus, the existing cooperative structure promises to make much more tractable the problems of winning wide acceptance of a single set of forest practices and standards. In 1996, a "Development Strategy for Sustainable Forestry and Regional Certification" was tested in the Birkaland region. Since the Nordics have a long history of forest management and essentially no remaining virgin forests, the implementation of a unified management approach that departs only slightly from existing practices appears to be feasible.

SWEDEN
Overview of Forest Situation

As with Finland, the area and stock of forest has increased substantially during most of the twentieth century. Between 1920 and 1990 Sweden's standing volume increased from 1,800 million cubic meters to 2,800 million cubic meters, or about 56%, reflecting a recovery from huge harvesting activities that occurred between 1850 and 1900. Swedish forests are fairly uniform in composition. The predominant species are Norway spruce and Scots pine, but birch and other hardwoods are also significant. See Table 4 for other quantitative characteristics of Swedish forestry.

Table 4. Summary Profile of Forestry in Sweden, 1994.

Total land area	44,996,000 hectares
Total forest area	28,950,000 hectares
Forests as percentage of total land area	64.3%
Productive, closed, nonreserved forest area (PCNR)	22,906,000 hectares
PCNR as percentage of forest area	79.1%
Forest ownership:	
Public	5,944,000 hectares (25.9%)
Forest industry	5,486,000 hectares (23.9%)
Other private	11,475,000 hectares (50.1%)
Value of selected forest product exports[1]	$9.0 billion
Value of selected pulp and paper exports[2]	$6.4 billion
Value of selected solid wood product exports[3]	$2.6 billion
Forest product exports as percentage of world exports	8.2%
Forest product exports as percentage of temperate forest exports	9.8

Notes on sources: 1. Based on 1994 FAO data for industrial roundwood, sawnwood, wood-based panels, paper and paperboard, wood pulp, and recovered paper.

2. Based on 1994 FAO data for paper and paperboard, wood pulp, and recovered paper.

3. Based on 1994 FAO data for industrial roundwood, sawnwood, and wood-based panels.

About 50% of Sweden's productive forest land is owned by small private ownerships and about 25% by forest industry firms. The area of natural forest is small, about 5% of the total, due to earlier pressures on the forest, primarily from agriculture. Fire has been largely eliminated since the nineteenth century, and during the 1980s about 800,000 hectares were declared natural reserves.

Forest Policy

Like in Finland, forest management has a long history in Sweden. Early reforestation laws date back to the sixteenth century. Laws governing forest activities have been periodically revised. A major forest law was enacted in 1979 and revised in 1994. In response to environmental concerns, the use of herbicides for weed control was abolished in 1980, with only limited exceptions. Under current law, the Swedish Forest Administration, within the Ministry of Agriculture, is responsible for the implementation of Swedish forest policy. The administration is composed of the National Board of Forestry and twenty-four county boards. The county boards are responsible for approving forest plans, enforcing forestry regulations, and providing extension services to private landowners.

The law requires landowners to submit plans detailing conservation measures to be taken at the time of harvest and restricts harvests in such a way as to regulate age class distribution. Large ownerships are subject to more rigorous regulation than small ownerships. Clearcutting is the prevalent harvesting system, with regeneration accomplished through the planting of pine and fir. Some seed tree regeneration for pine is also used. An increasing share of Sweden's harvest is from commercial thinnings.

Recent Changes in Laws and Regulations

In 1993, the Swedish Parliament approved a new forest policy based on the principles of sustainability and multiple forest use. The new Forest Act (1994) contains forest management provisions that must be observed on all forest land, irrespective of ownership. These include required regeneration, felling to promote regeneration, and the approval of felling plans by the Boards of Forestry. Felling is now subject to approval to ensure sustained yield and protect natural and social values of the forest.

This 1994 act liberalized some regulations but granted greater responsibility to the Forestry Boards for controlling environmental aspects of forest management. Landowners must submit detailed harvest plans and regenerate after cutting. A greater emphasis is given to biodiversity, being reflected on-the-ground by some modification in forestry practices and the designation of protected areas. Harvesting practices are becoming more environmentally benign, with fragile areas either not harvested or experiencing only a modified harvest. Environmental impact statements may be required, and the government can prohibit felling in some cases. Additionally, clearcuts are smaller and less common with "green tree retention" being practiced, in part, to provide seed trees for regeneration. The new law also prohibits the conversion of broadleaf forests of the south to conifer.

The shift to less intensive management and encouragement of more hardwoods is generally consistent with changing economic realities. Birch is now an important part of the feedstock of a pulp-mill, so it is no longer desirable, from an economic perspective, to try to eliminate birch from industrial forests. Furthermore, the costs of planting and artificial regeneration have increased, making clear the advantages to various forms of natural regeneration. Sweden, and Finland for that matter, would probably have moved in the direction of more diversity in their industrial forests and greater reliance on natural regeneration even if there had not been increased worldwide concern over sustainable forestry and biodiversity.

Sustainability Guidelines, Principles, and Certification

Sweden has been working with the World Wildlife Fund toward developing an FSC-approved certification program and has been part of the group working toward achieving a coordinated and harmonized Nordic Forest Certification system. At this point neither of these certification approaches has been fully accepted. Producers have encountered problems related to the logistics of certifying and tracking the chain-of-custody of timber from the large number of small owners. Many large industrial companies in Sweden have agreed to certification by FSC-accredited certifiers, and some have predicted that much of Sweden's forests will be covered by FSC accreditation within a short period of time (Hansen 1997). While the large forest owners have embraced "ecological landscape planning" involving long-term conservation strategies designed to mimic natural ecological processes and re-create a more diverse forest environment, the Swedish Federation of Forest Owners (a group of nonindustrial private landowners, or NIPFs), which has 88,000 members owning more than 5.7 million hectares of forest land, has decided not to participate in FSC certification.

Many in Sweden believe that they need to be in a position to "certify a very large part" of Swedish forest production and are working toward tying forest certification to a life cycle analysis. However, the limitation of certification to "a very large part" suggests the difficulty of achieving 100% certification given mixed forest ownership. As it stands, the inability to achieve broader participation in FSC forest certification, with the nonparticipation of the NIPFs, as well as chain-of-custody tracking problems, have stymied the process.

FRANCE

Overview of Forest Situation

The forest resources of France have been expanding since the late eighteenth century. An estimate of the forested area in 1890 was almost 50% higher than the area estimated in 1790, while the area of forest estimated for 1994 is again increased by about 50% (Ministry of Agriculture and Fisheries 1995b). Forest ownership is characterized by a large number of small owners, most of whom maintain forest land for purposes other than timber production. See Table 5 for other quantitative characteristics of French forestry.

Forest Policy

A major tool of France's forest policy is the Forest Code, which was first drawn up in 1827 and groups together all the laws and decrees that form the legal bases of forest policy. These policies are tied to a system of guidelines that are prepared at the regional level based on a consensus of foresters, landowners, and industry. Monitoring also takes place at the regional level, but violations are very rare. Almost all of the forests are privately owned. Nevertheless, fairly strict rules govern land use and forestry practices, with the state dictating practices to private forest owners.

Forest planning is required for forests greater than 25 hectares, with a lesser degree of planning for forest areas between 10 and 25 hectares. Although there is not a code of best practices as such, a

Table 5. Summary Profile of Forestry in France, 1994.

Total land area	57,153,000 hectares
Total forest area	14,810,000 hectares
Forests as percentage of total land area	25.9%
Productive, closed, nonreserved forest area (PCNR)	13,428,000 hectares
PCNR as percentage of forest area	90.7%
Forest ownership:	
Public	3,250,000 hectares (24.2%)
Forest industry	Included in Other Private (below)
Other private	10,178,000 hectares (75.8%)
Value of selected forest product exports[1]	$4.4 billion
Value of selected pulp and paper exports[2]	$3.2 billion
Value of selected solid wood product exports[3]	$1.2 billion
Forest product exports as percentage of world exports	3.9%
Forest product exports as percentage of temperate forest exports	4.7%

Notes on sources: 1. Based on 1994 FAO data for industrial roundwood, sawnwood, wood-based panels, paper and paperboard, wood pulp, and recovered paper.

2. Based on 1994 FAO data for paper and paperboard, wood pulp, and recovered paper.

3. Based on 1994 FAO data for industrial roundwood, sawnwood, and wood-based panels.

host of requirements are spelled out in guidelines called the "Red Book." Currently, a prime incentive for forest establishment is the low inheritance taxes for forest land. There are very few problems with violations of forest management regulations. If a forest is established and transferred via inheritance, the land cannot be converted back to agriculture use for thirty years. A tax on final forest products is used to fund forest extension activities and planting subsidies. The tax on land is decreased if a harvested forest is replanted.

Recent Changes in Laws and Regulations

Most forest regulations have changed only slightly over the past five to ten years, but a new Forestry Law is under development and is expected to be enacted this year. Proposed changes are being influenced at least in part by the "Greens," as well as by international developments including the Helsinki Process, in which France participated. The new law is expected to dictate a multiple use approach on all forests. Some changes have already been made in regional guidelines and regulations. The government is also establishing a series of small land areas for the conservation of genetic resources, to be overseen by scientific panels (Ministry of Agriculture and Fisheries 1995a). In addition, the European Community is expected to require some regulations related to forests, for example, that 15% of the forest land be set aside.

Sustainability Guidelines, Principles, and Certification

France, like most other countries, has a division of opinion about the efficacy of moving to a forestry certification regime. One view is that France has been practicing sustainable forestry as demonstrated by the many centuries, indeed millennia, during which forests have been sustained, utilized, and regenerated. In fact, the forest area in France has been expanding since at least 1800. The government has been a participant in the Helsinki Process and has developed publications detailing how France is looking at criteria and indicators.

GERMANY
Overview of Forest Situation

German forests cover about 10.7 million hectares, including forests that were formerly part of the German Democratic Republic (East Germany). Spruce/fir is the most prevalent forest type in Germany, followed by pine/larch and mixed hardwoods of beech and other species. Douglas fir and other nonnative species are also common. Ownership issues for appropriated land in eastern Germany will likely take many years to work out, but most of the forest land is being retained in government ownership, typically at the local level. State governments control about 33% of Germany's forest land, local jurisdictions control 20%, and some 750,000 individuals and other private entities own about 47%. The federal government is not a significant forest land owner, accounting for less than 2% of forest land, located on military and other government reservations. The majority of landowners do not depend solely on timber for income and tend to maintain forests for other reasons.

Although forests are locally governed, forest practices are fairly consistent throughout Germany. German forestry is characterized by uneven-aged management on rotations as long as 120 years for conifer and 160 years for hardwoods. As a result, inventories per hectare tend to be high. Harvesting is generally in the form of selection or shelterwood cuts. See Table 6 for other quantitative characteristics of German forestry.

Forest Policy

Germany has a long history of forest management. Currently, the Federal Forest Act of 1975 provides the general framework for forest management in Germany. Within that framework, it is the laws and regulations at the local level that are most influential and the thirteen "landers" (counties) and three cities that have the primary responsibility for forest regulation. Despite the high degree of local autonomy, forest practices are fairly consistent throughout Germany. The federal law was designed to provide a template for states and local jurisdictions by providing some parameters that are universally applied throughout the country, including a prohibition against clearcutting areas larger than 1 or 2 hectares. In recent years, an increased emphasis has been placed on hardwood silviculture. Management plans are required for forest holdings greater than 30 hectares, with stricter documentation requirements for owners of more than 70 hectares. The conversion of conifer stands to hardwood receives government support. Foresters employed by the landers or other local jurisdictions are responsible for forest management on the communal lands, and in many cases, on private lands as well.

The German government subsidizes front-end establishment costs for reforestation and provides annual payments to landown-

Table 6. Summary Profile of Forestry in Germany, 1994.

Total land area	35,700,000 hectares
Total forest area	10,740,000 hectares
Forests as percentage of total land area	30.0%
Productive, closed, nonreserved forest area (PCNR)	9,400,000 hectares
PCNR as percentage of forest area	87.5%

Forest ownership:

Public	3,168,000 hectares (33.7%)
Forest industry	1,861,000 hectares (19.8%)
Other private	4,371,000 hectares (46.5%)

Value of selected forest product exports[1]	$6.9 billion
Value of selected pulp and paper exports[2]	$5.4 billion
Value of selected solid wood product exports[3]	$1.5 billion
Forest product exports as percentage of world exports	6.3%
Forest product exports as percentage of temperate forest exports	7.5%

Notes on sources: 1. Based on 1994 FAO data for industrial roundwood, sawnwood, wood-based panels, paper and paperboard, wood pulp, and recovered paper.

2. Based on 1994 FAO data for paper and paperboard, wood pulp, and recovered paper.

3. Based on 1994 FAO data for industrial roundwood, sawnwood, and wood-based panels.

ers for up to twenty years. Subsidies are a third more generous for hardwood stand establishment than for conifer, and an articulated policy encourages more hardwood and mixed stands. Landowners are required to allow recreational use of forest land and have increasingly been subjected to harvesting restrictions.

Recent Changes in Laws and Regulations

A Nature Protection Law was recently enacted that gives broad authority to the Ministry of Environment for protecting natural areas and reducing environmental impacts from land uses. The law provides for new set-asides of public land to be left in natural condition. Five new national parks have been created in the eastern half of the country following unification, and one or more areas of hardwood forests are being considered for park designation. While there are no specific laws or regulations on threatened and endangered species, public lands are used to manage or introduce species that are declining. Some bird populations are protected during their nesting period, but landowners are not legally obligated to provide protection.

Sustainability Guidelines, Principles, and Certification

Environmental influence is pronounced in Germany and it is in Germany that the greatest market pressure for ecolabeling seems to be surfacing. Ironically, the German forest sector tends to discount the need to develop specific sustainability programs within Germany itself. Many landowners fear certification because of the difficulty it poses for tracking wood fiber from its source. The German view tends to be that sustainability is inherently the result of long rotations, selective harvesting, and very strict local regulation of forest practices. However, German forests tend to be of single-species composition and predominately conifer, which has raised biological diversity concerns in recent years. That conifer stands, which have been highly susceptible to wind damage in recent years, have provided an impetus for the government to encourage more hardwood growth.

Some communities have reached specific agreement with environmental organizations to take forest land out of production.

NEW ZEALAND
Overview of Forest Situation

Indigenous forests once covered 80% of New Zealand, but most of the forests were exploited and converted to pastureland during settlement by the Maori people and later by the Europeans. Native forests now cover about 24% of the total land area with plantations adding another 5%. Plantations began being established during the early part of the twentieth century as concerns about forest depletion grew. Radiata pine was found to grow extremely well under varied growing conditions and became the species of choice for intensive management. Over 90% of New Zealand's plantations are in radiata pine. Douglas fir and eucalyptus make up about 5% and 2%, respectively, with other species accounting for the balance. Planted production forests account for 99% of New Zealand round-wood production.

Currently, New Zealand has approximately 1.5 million hectares of forest plantations. Growth rates for radiata pine average 22 cubic meters per hectare per year on an average rotation of 27 years. Most of the country's remaining native forests are protected from commercial use by a combination of legislation and voluntary measures (Smith and Raison 1997). See Table 7 for other quantitative characteristics of forestry in New Zealand.

Table 7. Summary Profile of Forestry in New Zealand, 1994.

Total land area	26,868,000 hectares
Total forest area	7,900,000 hectares
Forests as percentage of total land area	29.4%
Productive, closed, nonreserved forest area (PCNR)	2,000,000 hectares
PCNR as percentage of forest area	25.3%
Forest ownership:	
Public	340,000 hectares (17.0%)
Forest industry	1,600,000 hectares (80.0%)
Other private	60,000 hectares (3.0%)
Value of selected forest product exports[1]	$1.4 billion
Value of selected pulp and paper exports[2]	$0.4 billion
Value of selected solid wood product exports[3]	$1.0 billion
Forest product exports as percentage of world exports	1.3%
Forest product exports as percentage of temperate forest exports	1.6%

Notes on sources: 1. Based on 1994 FAO data for industrial roundwood, sawnwood, wood-based panels, paper and paperboard, wood pulp, and recovered paper.

2. Based on 1994 FAO data for paper and paperboard, wood pulp, and recovered paper.

3. Based on 1994 FAO data for industrial roundwood, sawnwood, and wood-based panels.

Forest Policy in New Zealand

Major reform of natural resource policy was accomplished in the early 1990s, spurred by the economic and political changes that were transcending public policy in general and by the growing influence of environmental groups. The underlying forestry law dates back to the Forests Act of 1949, but it was amended in 1993. The amendments were designed to tighten regulation of activities on native forests, although commercial production of native species was, by then, fairly insignificant. It is widely accepted by all stakeholders in New Zealand that the goal of plantation management is timber production. Thus, lands planted in exotic species are under no requirement for multiple-use management (Smith and Raison 1997). However, management of plantations must conform to requirements of the Resource Management Act.

The Resource Management Act (RMA) was enacted in 1991 to replace more than fifty other statutes that had controlled most aspects of environmental protection and land use planning. The RMA is designed to be "effects based." That is, in theory, regulations imposed by the RMA are supposed to control specific and measurable problems associated with an activity, in contrast to being prescriptive in nature. The RMA is essentially a national land use planning law. It divides the country into forty-three regions and eighty planning districts, each with its own elected Resource Management Council, and establishes several categories of activities: permitted, controlled, discretionary, noncomplying, and prohibited.. Theoretically, any activity is permitted unless specifically forbidden by the regional council's plan. Councils may require that a resource consent be obtained for certain activities if it is concerned about impacts that can be linked directly to the activity. In general, forestry is not a permitted activity. Thus, most landowners must obtain resource consents for harvesting and, in some cases, planting. Any activity requiring the use or discharge of water requires a resource consent. Which forest activities require a resource consent depends on the region in the country; regional councils interpret the RMA differently. However, the regional agency is the only agency an operator must deal with for all environmental approvals, whether the activity involves the construction of a manufacturing plant or the harvesting of trees.

The RMA contains a general (and unenforceable) duty to avoid, remedy, or mitigate any adverse effect on the environment. Protection of endangered species is not specifically referenced or required under the statute. However, if a timber harvesting activity, for example, would place a species at risk, a federal agency or an environmental group could petition a court to issue an abatement notice preventing the activity. Usually, a consent decree will specify what actions would be necessary to mitigate any damage to habitat at risk.

Recent Changes in Laws and Regulations

New Zealand has undergone a remarkable economic transformation over the past decade. Having gone from a country with the world's fourth highest GDP in 1950 to a country strapped in debt and overly dependent on subsidized agriculture, New Zealand undertook radical economic reforms with the rise to power of the Labor Party in 1984. Agricultural subsidies were abolished almost overnight, and the government was decoupled from most other incursions into economic activity. Resource policy was also dramatically reformed. Privatization of the government forest operations began in the late 1980s and was essentially completed by 1995.

In 1993, the New Zealand forest products industry entered into a forest accord, under which it agreed to no longer conduct any harvesting activities in indigenous forests. In return, environmental groups agreed to promote plantation forestry as a sustainable land use. More recently, the industry has adopted management principles that set forth a philosophy and management guidelines that are being applied on industry lands.

Sustainability Guidelines, Principles, and Certification

New Zealand has supported all of the major international conventions and is a signatory to the Santiago Declaration (the Montreal Process). Forest certification, however, is being courted reluctantly and only in response to market pressures from Europe. The New Zealand industry is supporting the development of an ISO 14000 standard. The former executive director of the New Zealand Forest Landowners Association, now a member of the New Zealand parliament, chairs a drafting committee, which is writing a technical document for the forest sector within the ISO 14000 standard.

Some companies have been exploring FSC certification if only to remain players in the market niche that FSC seems to be nurturing among two or three large European buyers. FSC certification would require extensive documentation and audits but would not appreciably affect how most companies manage their land. Costs are estimated at $2.50 per hectare for ownerships over 100,000 hectares, and much higher for smaller units.

New Zealand forestry research organizations have redirected much of their research effort to address sustainability questions. One interesting project is looking at the presence of certain insect species as an indicator of changes to site conditions and, hence, as a potential measure of sustainability if insect populations can be tracked over time and associated with desirable conditions.

UNITED STATES

Overview of Forest Situation

Forests cover one-third of the land area of the United States, and 66% of those forests are classified as productive timberland. This level of natural productivity, combined with its large land base, allows the United States to be the leading producer of industrial roundwood and pulpwood in the world. More than 70% of America's forests are privately owned, either by the timber industry or by nonindustrial private landowners (NIPFs). The species composition, intensity of management, ownership patterns, and laws and regulations vary between states and regions.

Forest land in the United States is held by the U.S. Department of Agriculture's Forest Service, other federal and state government agencies, industry, and NIPFs in five main regions. The South has the highest percentage of the country's forest land with 29%. The North is a close second with 23%, followed by the Inland West (19%), and Alaska (18%). The Pacific Coast has the smallest forest area with 12%. However, the volumes of timber from Pacific Coast lands is high on a per land area basis. The South and North lead the country with 41% and 32% of the timber land, respectively. The Inland West (12.8%) and Pacific Coast (11.2%) are next, while Alaska's 3% share is the smallest.

Nationally, the Forest Service owns 17% of all forest land, with an additional 10% held by other public institutions. The remainder is controlled by private owners, who are divided into two categories: industry, whose primary objective is timber management, and NIPFs whose goals may or may not be geared toward wood and fiber production. Industrial owners hold 14% of all forest lands; NIPFs own 59%. There are variations in this ownership pattern by region. Greater proportions of private forest land occur in the South and North. National forest and other public owners dominate in the West and Alaska. Most of the timberland area in the North and the South is held by NIPF owners (71% and 70%, respectively). Forest industry holds less than 22% of the timberland in any region, with the greatest percentage in the South. Most of the productive timberland in the Inland West (67%), Pacific Coast (53%), and Alaska (60%) is owned by federal or state governments.

The area of productive, reserved forest lands that have the potential to be commercial timberland but have been removed from timber production by legislation amount to about 7% of the total forest base in the United States. Greater amounts are reserved in the Inland West (16%), Pacific Coast (11%), and Alaska (29%) than in the North (5%) and South (2%).

Softwood volumes are distributed across the country, with 65% of softwood stocks in the West and the remainder in the East. Public lands contain over half of the softwood sawtimber volume and most of the remaining old-growth forest. Hardwoods predominate in the North and South, with only a minor share of the hardwood harvest occurring in the West. The South leads the country in timber removals with almost 55% of the U.S. total. Nearly half (49%) of all timber removals occur on lands owned by NIPFs. Industrial properties provide 33% of the national harvest. National Forests and other public sources account for 15% and 6%, respectively.

See Table 8 for other quantitative characteristics of U.S. forestry.

Forest Policy

American attitudes and policies toward forests have gone through a number of phases since the settlement of the country. During the colonial period, laws regulating forest use were passed by the English monarchy to reserve individual trees and species of trees for use by the British Navy. The period after independence to the late 1800s and early 1900s was dominated by the dispersal of federal lands to private owners. Large areas of forest were converted to agricultural uses while thousands of acres were harvested and abandoned by the lumber barons.

The late nineteenth and early twentieth centuries saw the creation of the national forests, wildlife refuges, and parks. Legislation in this period was limited mostly to organic statutes that allowed

Table 8. Summary Profile of Forestry in the United States, 1994.

Total land area	915,941,000 hectares
Total forest area	298,135,000 hectares
Forests as percentage of total land area	32.6%
Productive, closed, nonreserved forest area (PCNR)	198,123,000 hectares
PCNR as percentage of forest area	66.5%
Forest ownership:	
Public	27%
Forest industry	14%
Other private	59%
Value of selected forest product exports	$14.0 billion
Value of selected pulp and paper exports	$ 8.3 billion
Value of selected solid wood product exports	$ 5.7 billion
Forest product exports as percentage of world exports	12.7%
Forest product exports as percentage of temperate forest exports	15.2%

for the establishment, acquisition, and administration of federal reserves and preserves. Another phase coincided with the creation of forestry schools and state forestry agencies in the United States, which, in turn, contributed to better management as cut-over lands were replanted and destructive wildfires controlled. This general trend of stewardship continued into the 1960s and was a period that saw great advances in the science and practice of forestry. The optimism prevalent in this period is evident in the Multiple-Use and Sustained Yield Act of 1960 that stipulated that the National Forests would (and could) simultaneously provide recreation, range, and timber, while protecting watersheds, fish, and wildlife.

Public dissatisfaction with environmental conditions grew during the 1960s and ushered in new attitudes and legislation regarding forests. The decade saw the passage of the Wilderness Act of 1964 and two early attempts to protect endangered wildlife, and ended with the passage of the National Environmental Policy Act (NEPA) of 1969. In many ways, these acts created a precedent for restoration, rehabilitation, protection, and revival of the country's lands, waters, and wildlife through federal legislation. Many of the most important and far-reaching federal environmental laws were passed in the 1970s: the Clean Air Act of 1970, the Amendments to the Federal Water Pollution Control Act (Clean Water Act) of 1972, the Federal Environmental Pesticide Control Act of 1972, the Endangered Species Act of 1973, the Forest and Rangeland Renewable Resources Planning Act (RPA) of 1974, and the National Forest Management Act (NFMA) of 1976.

Few of these laws were enacted specifically to control forestry operations, but almost all contain elements that govern planning or conduct of some forestry activities as part of their broad environmental protection mandates. Although the RPA and NFMA directly affected forest lands, their effectiveness was limited since they were largely applicable only to the national forests. Of greater impact were the Clean Water Act, which specifically listed controlling certain forestry-related activities on private lands, and the Endangered Species Act, which provided protection for threatened wildlife regardless of whether such species were on private or public property.

In 1989, for example, the northern spotted owl was listed as threatened by the U.S. Fish and Wildlife Service. One result of the listing has been the reduction in softwood timber removals in the Pacific Northwest, where the spotted owl is found, and a shift in production to the U.S. South and Canada. Other environmental laws have affected how forestry is practiced in the United States, and because these statutes were administered via state implementation programs, state governments have evolved into the primary regulators of American forests.

Partly to avoid stricter federal legislation, many states implemented voluntary, quasi-voluntary, and regulatory *best management practices* (BMPs) in the 1970s and 1980s that set guidelines for water quality, replanting, and road building. Some other states have passed additional regulations that affect forest practices on private lands. States continue to draft, revise, and strengthen BMPs and have the greatest ability to affect practices on the large area of private forests in the United States.

In addition to the protective regulatory approach exhibited in environmental and forestry laws, some incentive programs were passed to encourage better stewardship of nonindustrial private lands. A rider to the Agricultural Conservation Program in 1973 created the Forestry Incentives Program, which authorized cost-share payments for reforestation and timber stand improvements on private lands. A similar program affecting agricultural lands, the Conservation Reserve Program, was passed to foster the conversion of marginal crop lands to forests.

Recent Changes in Laws and Regulations

In recent years, U.S. forest policy has been shaped mostly through judicial interpretation of existing environmental laws. Private citizens and government bureaus have challenged numerous aspects of environmental statutes in the courts. Some cases have made their way to the Supreme Court. Additionally, several major statutes are past due for reauthorization. In addition, new regulations are being drafted for the management of the National Forests and new National Forest legislation is under consideration by the Congress.

However, given the lack of consensus on the direction the country should take regarding the future of environmental and forest regulation, this period of judicial interpretation may well continue into the next decade.

Sustainability Guidelines, Principles, and Certification

Almost three-quarters (73%) of America's timberlands are controlled by private owners (both industry and NIPFs), and those lands produce 82% of the domestic wood supply. Furthermore, the timber industry, while owning only 14% of all timberlands, harvests 33% of the country's wood. While some states have guidelines for timber harvests and management on private lands, federal government involvement on private land has been less direct.

The United States has been an active participant in the Montreal Process and the Santiago Declaration. President Bill Clinton has expressed a commitment to execute the criteria and indicators expressed in the Santiago Declaration by the year 2000. To meet that goal, the U.S. Forest Service has been designated the lead federal agency for the process. The Forest Service has completed two examinations to determine how to quantify the objectives defined by the criteria and indicators, and has further identified areas where data either do not exist or need to be collected.

State efforts toward sustainable forestry are nearly as diverse as the states themselves. Actions can be separated into three broad categories: task-force actions and reports, new legislation, and certification of state lands by third-party certification organizations.

North Carolina, Pennsylvania, Minnesota, and Maine have completed a review of state practices by blue ribbon panels appointed by their respective state governors. In general, these panels have produced a set of nonbinding guidelines and recommendations of how public and private landowners within each state can improve the stewardship of local forests. North Carolina's efforts have called for the creation of the Southern Center for Sustainable Forestry that will serve to conduct research and extension education in the South.

No laws have passed state legislatures specifically addressing sustainable forestry. Connecticut is finalizing its BMP regulations, while West Virginia and Virginia are currently considering passing state forestry guidelines. Three initiatives have failed in the past year: in Tennessee and Kentucky the state legislatures both failed to pass BMP-type rules for forest practices; the citizens of Maine narrowly rejected a measure to ban all clearcutting in the state, but not by a great enough majority to completely defeat the proposal. Finally, Pennsylvania and Wisconsin have contracted with private third-party certification agencies to examine state management practices and make recommendations regarding improving the sustainability of current practices. At the present time it is unclear whether such a review will lead to certification of state management practices and the timber harvested from state forests.

Private action for sustainable forestry can be separated into two types: independent third-party certification organizations and a voluntary industry initiative. Two main certification organizations are active in the United States at the present time. Both are accredited by the Forest Stewardship Council (FSC) and abide by the sustainable forestry guidelines promulgated by the FSC. Scientific Certification Systems certifies forestry operations through its Forest Certification Program, while Rainforest Alliance works with a group of regional cooperators to certify operations and award them its Smart Wood logo. Both groups have certified a small number of forests in the United States and other countries.

The voluntary industry program is the American Forest and Paper Association's (AF&PA) Sustainable Forestry Initiative (SFI). Developed in 1994, SFI is a set of standards for silvicultural practices that member companies must agree to implement to maintain standing in the AF&PA. Despite losing some members, the AF&PA has held firm to SFI and is now sponsoring logger education and other programs designed to affect positive changes in the way private lands are managed.

Appendix B
Criterion 7

Coming out of the Santiago Declaration in early 1995 was a set of criteria and indicators for the conservation and sustainable management of temperate and boreal forests (see the discussion on criteria and indicators in Chapter 3 of this book). These were intended to provide a common understanding of what is meant by sustainable forest management and provide a common framework for describing, assessing, and evaluating a country's progress toward sustainability *at the national level*. Thus, the criteria and indicators help provide an international reference for policymakers in the formulation of national policies and a basis for international cooperation aimed at supporting sustainable forest management.

Related to the overall policy framework that is designed to facilitate the conservation and management of forests of a country is Criterion 7 and its associated indicators. Criterion 7 includes the legal, institutional, and economic framework for forest conservation and sustainable management. Indicators of these criteria include the following:

- Well-defined property rights, the extent of periodic forest-related planning, the use of best-practice codes for forest management
- Provision of management for the conservation of special environmental cultural, social and scientific values
- Adequate infrastructure to facilitate the support of forest products and management
- Enforcement of laws, regulations, and guidelines

- Investment and tax policies, nondiscriminatory trade policies for forest products
- Ability to measure and monitor changes in conservation and sustained management
- Capacity to conduct and apply research

One issue to be addressed in this study was an investigation as to whether the eight selected countries possess an institutional framework that can translate laws and policies into forest conservation and sustainable management at the national level.

The eight countries examined in this study included many that have among the most well-developed legal, economic, and social institutions in the world. Four are in western or central Europe, two in North America, and one in Oceania. All these countries have long histories, although not always uninterrupted, during which institutions developed. These include legal systems and the rule of law, as well as social and economic institutions. All have well-developed educational systems with almost universal literacy and democratic-style governments which provide for "public involvement activities." Thus, these seven countries appear to have the desired prerequisite institutions well established.

One of the eight countries is in South America and has experienced substantial economic growth but is just arriving into the family of the developed countries. In many respects it might still be characterized as a developing country. This country has had a much briefer period in which to develop legal, economic, and social institutions with which to ensure the smooth transfer of national laws and policy decisions through the system. Furthermore, it has lower literacy rates, a much shorter national history as a functioning democracy, and so forth. Nevertheless, this country has made substantial economic and social progress in recent decades and offers the potential of an institutional framework that can successfully implement laws and policies throughout the system.

Of the eight selected countries, all but the United States have instituted major new forest laws or national policies within the past few years (see the case studies in Appendix A)—mostly since the 1992 Earth Summit. These countries appear to have anticipated the evolving reality that includes sustainable forestry and have attempted to update their laws and policies accordingly. In the United States, although new laws have not been enacted, forest management has been affected by recent litigation and court interpretations of existing statutes. For example, the Endangered Species Act, as now interpreted, provides substantial protection of threatened or endangered species and includes provisions regarding habitat on both private and public lands.

Table 1 briefly examines and assesses the "inadequacy of official laws, policies and regulations in light of the actual performance of the country's forests" through the prism of Criterion 7. Additional relevant material is found in Chapter 5 of this book.

Table 1. Criterion 7: An Assessment of Legal, Institutional, and Economic Elements Relating to Sustainable Forestry.

Country	Legal, institutional, and economic framework	Examples of indicators	Ability of institutional framework to translate national policies into performance	Evidence of changes in performance
Canada	Well developed	Property rights, forest planning, infrastructure	High	New harvesting practices in British Columbia
Chile	Moderately developed	Property rights, forest tax policies	Medium	Reforestation activity related to the tax law
Finland	Well developed	Property rights, infrastructure, forest practices codes	High	Changing harvesting practices to enhance biodiversity
Sweden	Well developed	Property rights, forest research, forestry codes	High	Practices encourage "green" tree retention, multiple-species stands
France	Well developed	Regional forest practices codes	High	More attention to biodiversity
Germany	Well developed	Forest plans, property rights, infrastructure	High	New forest park set-asides
New Zealand	Well developed	Education, property rights, infrastructure	High	Protection of native forests
United States	Well developed	Education, research, infrastructure	High	Set-aside of old-growth forests

References

Abt, Robert, and Brian Murray. 1997. Personal conversation, reporting on work in progress.

Binkley, Clark S. 1995. Designating an Effective Forest Research Strategy for Canada. *Forestry Chronicle* 71: 589–95.

Binkley, Clark S., Michael Perky, William A. Thompson, and Ilan B. Vertinsky. 1994. A General Equilibrium Analysis of the Economic Impact of a Reduction in Harvest Levels in British Columbia. *The Forestry Chronicle* 70(4):449–54.

Bourke, I. J. 1996. *Global Trends in Marketing Environmentally Certified Forest Products.* Paper delivered to the Australian Outlook Conference, 6–8 February. Canberra, Australia.

CCFM (Canadian Council of Forest Ministers). 1995. Defining Sustainable Forest Management: A Canadian Approach to Criteria and Indicators. Ottawa, Canada: Canadian Forest Service, Natural Resources Canada.

CSA (Canadian Standards Association). 1996a. *A Sustainable Forest Management System: Guidance Document.* Z808. Draft: February 6.

———. 1996b. *A Sustainable Forest Management System: Specification Document.* Z809. Draft: February 6.

Drake, David. 1995. *Background Report on Global Movement Towards Sustainability for Canada/U.S. Softwood Lumber Consultation.*

Dunsworth, B.G., and S.M. Northway. 1996. *Modeling the Impact of Harvest Schedules on Biodiversity: 1995 Progress Report.* Forest Renewal British Columbia Research Report; Lands Resources and Environment, Agreement Ref. No. (SCBC) Fr-95-96-192.

Elliott, C., and R.Z. Donovan. 1997. Introduction. In Viana and others 1996.

Flasche, F. 1997. Presentation to the II International Forest Policy Forum, Solsona, Spain, March 12.

FAO (Food and Agricultural Organization of the United Nations). 1987. *Tropical Forestry Action Plan.* Prepared by FAO, the World Bank, World Resource Institute and the UNDP, Rome.

FSC (Forest Stewardship Council). 1997. *Pacific Coast Guidelines Working Group Regional Standards Draft 2.* September.

Ghazalali, B.H., and M. Simula. *Certification Schemes for All Timber and Timber Products.* Prepared for the International Tropical Timber Organization, Yokohama, Japan. April 15.

Gluck, P. R., Tarasofsky, N. Byron, and I. Tikkanen. 1997. *Options for Strengthening the International Legal Regime for Forests.* A report prepared for the European Commission under study contract B7-8110/96/000221/D4.

Greene, John L. and others. 1995. *The Status and Impact of State and Local Regulation on Private Timber Supply.* USDA Forest Service, General Technical Report RM-255. Washington, D.C.: U.S. GPO.

Grime, J.P. 1997. Biodiversity and Ecosystem Function: The Debate Deepens. *Science* 29 (August): 1260–61.

Groves, Michael., Frank Miller, and Richard Donovan. 1997. Chain of Custody. In Viana and others 1996.

Haley, David. 1996. Paying the Piper: The Cost of the British Columbia Forest Practices Code. Paper presented at the conference, Working with the British Columbia Forest Practices Code, and Insight Information Inc. Sponsored by the *Globe and Mail*, Vancouver, British Columbia, April 15–17.

Hansen, Eric. 1997. Forest Certification and Marketing, Forest *Products Journal* 47(3): 16-22.

Heaton, K., and R. Donovan. 1997. Forest Assessment. In Viana and others 1996.

Kajanus, Miika, and Harri Karjalainen. 1996. WWF's Ecolabelling Project: Costs of Ecolabelled Forestry. Paper presented 20 March to the meeting of the Scandinavian Forest Economics, Joensuu Finland.

Kiekens, J.P. 1995. Timber Certification: A Critique. *Unasylva*, 46(184): 27.

Korotov, A.V., and T.J. Peck. 1993. Forest Resources of the Industrial Countries: An ECE/FAO Assessment. *Unasylva* 44( ):.

Kuusela, Kullervo. 1994. *Forest Resources in Europe 1950–90.* Cambridge: Cambridge University Press.

Kurz, W.A., and M.J. Apps. 1995. An Analysis of Future Carbon Budgets of the Canadian Forest. In M.J. Apps, D.T. Price and J.Wisniewski (eds.), *Boreal Forests and Global Change.* Dordrecht: Kluwer Academic Press.

Lugo, A.E., J.A. Parrotta, and S. Brown. 1993. Loss in Species Caused by Tropical Deforestation and Their Recovery Through Management. *Ambio* (22): 106–9.

Menzies, Nicolas. 1992. Sources of Demand and Cycles of Logging in Premodern China. In J. Darcavel and R.P. Tucker (eds.), *Changing Pacific Forests: Historical Perspectives on the Forest Economy of the Pacific Basin*. Durham, North Carolina: Duke University Press.

Michaelis, Lynn O. 1996. Challenges to Private Land Management in the Pacific Northwest: Is There a Future? Presented to the conference, Forest Policy: Ready for Renaissance, at Olympic Natural Resource Center in Forks, Washington. September.

Ministry of Agriculture and Fisheries. 1995a. *Forestry Policy in France*. Paris, France: Countryside and Forest Department.

———. 1995b. *Sustained Forest Management in France*. Paris, France: Countryside and Forest Department.

NFC (Nordic Forest Certification). 1996. *Report No. 1.* February.

Ozanne, L.K. and R. P. Vorsky. 1996. Willingness to Pay for Environmentally Certified Wood Products: The Consumer Perspective. Forthcoming in *Forest Products Journal*.

Palo, Matti. 1995. Continuity and Change in Finnish Forest Sciences. In *Research in Forest and Forest Science in Finland*. Helsinki: Helsinki University Printing House.

Rotherham, Terry. 1996. Forest Management Certification: Objectives, International Background and the Canadian Program. *The Forestry Chronicle* 72(3).

Sedjo, R.A. 1992. Forest Ecosystems in the Global Carbon Cycle. *Ambio* 21(4): 274–77 (4 June).

———. 1995. Forests: Conflicting Signals. In R. Bailey (ed.). *The True State of the Planet*. New York: Free Press.

Sedjo, R.A. and D. Botkin. 1997. Using Plantations to Spare Natural Forests. *Environment* 39(10; December): 14–30.

Simula, Markku. 1997. Economics of Certification. In Viana and others 1996.

Smith, C.T. and Raison, R.J. 1997. *The Utility of Montreal Process Indicators for Soil Conservation in Native Forests and Plantations.*

Sutton, Wink. 1996. An Unforeseen Aspect of Certification of Wood Origin. *The Sutton Weekly Comment*. 27 September.

Taylor, Michael. 1996. *Implementation of Sustainable Forest Management Systems and Forest Management Certification in North America*. October. Portland, Oregon: World Forest Institute.

Tikkanen, I. 1997. Personal conversation.

Upton, Christopher, and Stephen Bass. 1996. *The Forest Certification Handbook*. Delay Beach, Florida: St. Lucie Press.

van Kooten, G.C. 1994. *Cost-Benefit Analysis of BC's Proposed Forest Practices Code*. Vancouver: Forest Economics and Policy Analysis Research Unit of British Columbia

Viana, Virgilio. 1997. Certification as a Catalyst for Change. In Viana and others 1996.

Viana, Virgilio, M. Jamison Ervin, Richard Z. Donovan, Chris Elliott and Henry Gholz (eds). 1996. *Certification of Forest Products: Issues and Perspectives*. Washington, D.C.: Island Press.

WCED (World Commission on Environment and Development). 1987. *Our Common Future*. Oxford: Oxford University Press.

WCFSD (World Commission on Forests and Sustainable Development). 1997. *Forest Issues Brief: First Discussion Draft*. North American Regional Public Hearing, Winnipeg, Manitoba, Canada. 29 September–5 October, 1996. Published 3 July. Geneva: WCFSD Secretariat.

Williams, M. 1988. The Death and Rebirth of the American Forest: Clearing and Reversion in the United States, 1900–1980. In *World Deforestation in the Twentieth Century*, edited by John F. Richards and Richard P. Tucker. Durham: Duke University Press.